ENCYCLOPEDIA OF ORGANIC CHEMISTRY

ENCYCLOPEDIA OF ORGANIC CHEMISTRY

Vol. 7

By

Sananda Chatterjee

DAE, DCA, CMCNet

Scientific Consultant

Institute for Natural Sciences

Kolkata

(West Bengal)

D P H

DISCOVERY PUBLISHING HOUSE PVT. LTD.

NEW DELHI-110 002

Published by:
Tilak Wasan
DISCOVERY PUBLISHING HOUSE PVT. LTD.
4383/4B, Ansari Road, Darya Ganj
New Delhi-110 002 (India)
Phone : +91-11-23279245, 43596064-65
Fax : +91-11-23253475
E-mail : parul.wasan@gmail.com
discoverypublishinghouse@gmail.com
web : www.discoverypublishinggroup.com

***First Edition:* 2012**

ISBN: 978-81-8356-875-3 (Set)

Encyclopedia of Organic Chemistry

Printed at:
Shree Balaji Art Press
Delhi

Preface

Organic Chemistry is that branch of chemistry, which deals with the reactions related to the life form, or one of the source of the life form, carbon (we can say the end of it).

This carbon mainly conjugates with different elements such as nitrogen, hydrogen and oxygen sometimes metals and other non-metals also). Organic chemistry studies mainly the extraction, structure, properties, of these compounds.

In this book chapters deals with the different methods of purification, extraction of compounds, some processes were used and devised by Alchemists.

The extraction preservation of natural dyes are examples of alchemical work, these been discussed in the chapter "dyes indicators and pigments".

Dyes is the mal-pronunciation of the damathi word dyum meaning colour, it was, the ancient Egyptians who use different natural pigments to extract different colours.

In this book the organic compounds has been mainly classified into two broad groups, the aromatic and the aliphatic compounds. The extraction of alcohol is definitely an alchemical process'.

The destructive distillation of wood is one of the finest processes devised by the alchemists for the preparation of methanol is one of the important topics of this book.

Looking ahead in to the modem world, the world of future, the subject comes in vision is the biochemistry, the base of which is organic chemistry, the chapter containing carbohydrates, and the chapter, chemical compounds of biological and biochemical interest is the best element of focus for the students stepping foot in the advanced staircase of biochemistry.

The last chapter, chapter 41, discusses the different organic reactions and their suitably mechanisms, this will help those students who are very much afraid of organic formulas and reactions. This chapter also deals with the chirality one of the interest of Dr. Palash Gangopadhyay (author's friend).

Organic Chemistry is not a mere subject, it is the rhythm of life tuned in the essence of carbon, which looks black but carries the dazzleness of diamond in its heart.

SANANDA CHATTERJEE

Contents

CHAPTER

35

Heterocyclic Compounds

Introduction

It has already been noted that there are different types of aromatic closed ring compounds, these compounds have a slight, difference in there *closed rings*; there are compounds of other elements addition to carbon. Therefore they are enlisted under ***Heterocyclic compounds***. Owing to the close relationship to the members of the aliphatic series, certain derivatives of these have been discussed in aliphatic series. These are readily prepared from open chain compounds, and are easily regenerated. The ring system of the compounds about to discussed are distinguished by greater stability i.e. they are less readily ruptured.

Most of such rings resemble the benzene nucleus in containing several unsaturated linkage, and heterocyclic compounds also possess many points in common with those of benzene series.

Heterocyclic systems are known in great variety and their study forms one of the most interesting branches of organic chemistry. Only derivatives of the most interesting branches of organic chemistry.

Only derivatives containing sulphur nitrogen and oxygen will be considered in this chapter.

Compounds of that type where these nonmetallic ions are replaced by metallic or metalloid elements, these are called chellates and will be discussed in chapter 38 organo metallic chemistry and compounds.

As in the case of carbocyclic compounds a distinction is again drawn between rings containing three, four, five six and still higher number of atoms. The termed hetero – atoms and according to the number of those present we shall discuss of mono di or tri heterocyclic rings and so on.

In connection with the various carbon rings it has been explained in the chapter of organic mechanism that the five and six membered types are relatively unstable as is shown by the fact that they are difficult to form and readily break up again. Those containing five and six – membered rings, on the otherhand, are usually distinguished by comparatively highly stable. It should be well noted that the number of heterocyclic systems is increased still further by the existence of condensed pole nuclear types. Just as naphthalene is composed of two benzene nuclei and phenanthrene of benzene and a naphthalene nucleus, so in the same manner benzene naphthalene and other rings may condense with heterocyclic systems. A complicated example of this has already been met with indanthrene and numerous others will be found in connection with quinoline indole and their derivatives.

Special importance attaches to those compounds in which a five or six membered ring containing nitrogen is present. This class includes the vegetable alkaloids and antipyrin, of great value in medicine and dye – stuff as indigo. Compounds derived from these system will therefore be treated in greater detail. Five membered ring containing two or more atoms of nitrogen are frequently named with the ending *azole* (pyrazole, triazole) and six membered rings with the ending *azine* (pyrazine, tri azine tetrazine)

Pyrrole Furane and Thiophene Groups

The heterocyclic compounds pyrrole furane and thiophene which stand in close relationship to each other will be described first.

Pyrrole Group

Among five – membered rings systems containing nitrogen the pyrrole group stands out automatically prominently. Included under this heading are all those chemical compounds, the molecules of which contain a ring built up of four carbon atoms and a nitrogen atom.

+NH_3 → ; −H_2O →

ethyl 3,4-dioxopentanoate
Diaectosuccinic ester

Z
diethyl (3)-2-acetyl-3-(1-aminoethylidene) succinate

diethyl 2,5-dimethyl-1H-pyrrole-3,4-dicarboxylate

The presence pf this ring has been established in a series of important vegetative bases, which where originally regarded solely as derivative of the six – membered ring compound pyride, viz nicotione, hygrine custhygrine, stropine, hyoscyanine cocaine, tropacocaine and others. Further, Fisher obtained pyrrolidine – 2 – carboxylic acid as a hydrolytic product of various proteins and other investigatiors including Willstätter proved that hæmoglobin and chlorophyll are pyrrole derivatives, thus revealing an interesting connection between the colouring matter of blood and leaves.

The above examples provide sufficient illustration of the importance of pyrrole derivatives.

PYRROLE

Introduction

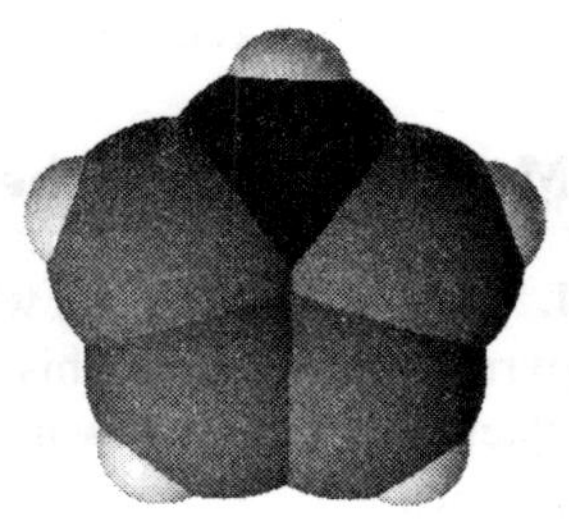

Pyrrole itself the parent substance of this class, was first discovered in coal tar and bone tar, and is also present among the distillation products of bituminous shale. Bayer was the first to advance the formula (II) now generally accepted for pyrrole.

The structural resemblance between pyrrole compounds and those of furane and thiophene has been clearly demonstrated by the work of L Knorr and Parr, on the formation of these compounds from γ - diketones or their enolic modifications. This has been described below.

Nomenclature of Pyrrole and its Derivatives

The positive of substituent in the pyrrole nucleus is usually indicated by the numbers 1 to 5 as in formula III.

(III) H N 1; HC 5, 2 CH; 4, 3; CH – CH

(IV) H N; HC α', α CH; β', β; CH – CH

Another system makes use of the letters a, b as in (IV). Since positions α and α' are equivalent, and also positions β and β', it is often convenient to distinguish monosubstitution products as α and β compounds respectively. Derivatives containing a substituent attached to nitrogen are frequently described as N – compounds.

From the above it is seen that each C – monosubstitution products of pyrrole can exist in two compounds as an a or β - derivative. Each C – distribution product can occur in four modifications viz as in αα" - αβ' or ββ' – derivative.

Dihydro – pyrroles as *pyrrolides* Keto pyrrolines are termed as *pyrrolones* and the keto – *pyrrolidines* are described as *pyrrolidones*

Distinctions are also drawn between various pyrrolones and pyrrolidones, according to the position and number of keto – groups in the molecule. The term "pyrrolidine" is commonly used to describe 2

1*H*-pyrrole

2,5-dihydro-1 *H*-pyrrole
pyrroline

pyrrolidine

pyrrolidin-2-one
pyrrolidone

– keto pyrrolidine or as α - pyrrolidone derivatives and may be regarded as lactams of g - amino acids. The imides of the succinic acid group, of which succinimide itself is the simplest representative, are αα' or 2: 5 diketo – pyrrolidines.

Methods of Synthesis of Reduced Pyrrole Derivatives

I. The 1 – 4 diketones, when treated with ammonia or primary amines are transformed with ease into pyrrole derivatives. This synthesis is affected with equal readiness when the reagents are dissolved in glacial acetic acid, water or ether and appears to depend on the intermediate formation of amino ketones.

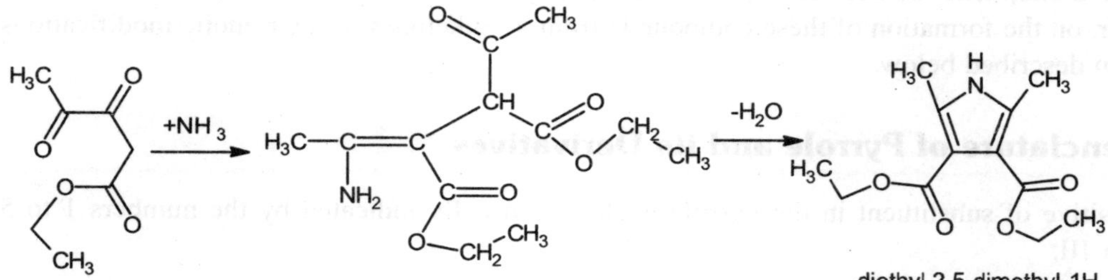

+NH_3 → ; $-H_2O$ →

Diaectosuccinic ester

diethyl (3Z)-2-acetyl-3-(1-aminoethylidene) succinate

diethyl 2,5-dimethyl-1H -pyrrole-3,4-dicarboxylate

This reaction has provided of service in the preparation of a great number of pyrrole derivative. Any γ - diketo – compound of the may be employed and the place of ammonia may be taken by primary amines, amino acids, hydroxylamine or phenyl hydrazine.

II. A second synthesis of pyrrole is also due to L Knorr, who succeeded in preparing 2: 4 dimethyl pyrrole 3:5 dicarboxylic ester by reducing an equimolecular mixture of iso – nitroso acetoacetic ester by means of zinc dust and glacial acetic acid. In a similar manner other pyrrole derivatives were prepared by reducing mixtures of esters of β - ketonic acids and their isonitroso compounds.

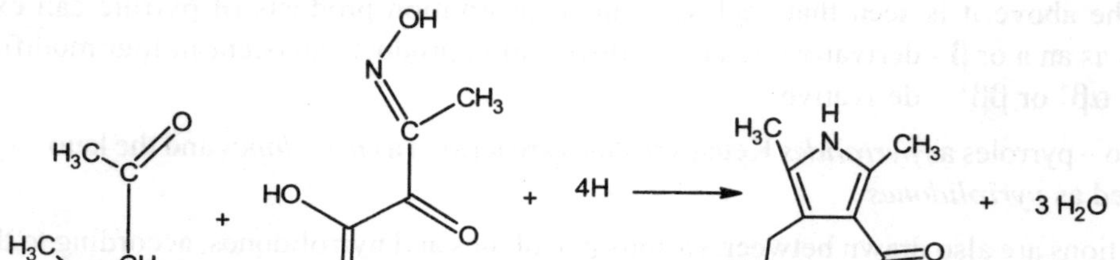

+ + 4H → + 3 H_2O

pentan-2-one

4-ethyl-2,5-dimethyl-1*H*-pyrrole-3-carboxylic acid

(3*E*)-3-(hydroxyimino)-2-oxobutanoic acid

3. In certain cases the reduction of a mixture of an amino acid ester (e.g. aminocrotonic ester ammonium chrotonate) with a 1:2 diketone also leads to the formation of derivative of pyrrole. Reduction may be dispensed with altogether if the diketone is replaced by 1:2 keto alcohol of the type of benzoin.

3-hydroxybutan-2-one + methyl (2Z)-3-aminobut-2-enoate → methyl 2,4,5-trimethyl-1H-pyrrole-3-carboxylate + $2H_2O$

4. A method first used by Hantzsch is to treat β - ketone esters with chloro – acetone and ammonia, when pyrrole carboxylic esters are formed.

5. Pyrrole can also be obtained from glutamic acid and also its calcium salt

Glutamic acid → pyrrolidine-2-carboxylic acid → 1H-pyrrole

6. Pyrrole is also formed by heating to dull redness a mixture of ammonia and acetylene in an iron tube.

NH_3 + CH≡CH → 1H-pyrrole
ammonia, acetylene

Physical Properties

Molecular Formula	= C_4H_5N
Formula Weight	= 67.0892
Composition	= C(71.61%) H(7.51%) N(20.88%)
Melting Point	= –23 °C
Boiling Point	= 129-131 °C
Molar Refractivity	= 20.68 cm^3
Molar Volume	= 67.7 cm^3

Parachor	= 166.8 cm^3
Index of Refraction	= 1.522
Surface Tension	= 36.7 dyne/cm
Density	= 0.990 g/cm^3
Log P	= 0.75
Polarizability	= 8.20 cm^3
Monoisotopic Mass	= 67.042199 Da
Nominal Mass	= 67 Da
Average Mass	= 67.0892 Da

Chemical Properties

It is insoluble in aqueous alkalies and slowly dissolves in acids. On long standing or warming in acid solution a red flocculent* precipitate of a substance precipitates which is called pyrrole red.

In pyrrole vapour a pine splint moisten with hydrochloric acid is coloured a pale red, which rapidly changes to an intense caramine red this reaction is employed as the identification of pyrrole.

In the presence of dilute acids pyrrole readily unites with a number of compounds containing the group —CO —CO (such as phenyl glyoxalic acid, benzil, phenanthraquinone, and alloxan) with the formation of dye stuffs.

Salt Formation with Pyrrole

Pyrrole is a very weak base which dissolves slowly in dilute acids, with strong acids it is rapidly resinified.

Even from solutions in dilute acids it is only possible to isolate definite simple salts in a few cases, and resinification readily takes place. This action of acids on pyrrole is possibly due to polymerization.

Pyrrole combines with picric acid to give a very unstable picrate with certain metallic salts it yields double compounds.

Salt formation can only be estimated definitely with derivatives of pyrrole which are stable towards strong acids. Chief among these are derivatives of dimethyl pyrrole containing acetyl or estimated carboxylic acid groups. The negative radicals make the ring more resistant and at the same time the methyl groups increase the basic properties. In its power of forming salts, pyrrole can be compared with diphenylamine.

Pyrroles are aromatic in character and possess points common with both phenols and aromatic amines, as may readily been seen from their reactions.

---

*A spongy floating mass.

The analogy with phenols is shown by the similarity in behaviour of the :NH group in pyrrole with that of the phenolic hydroxyl group and formation of a solid potassium compound C_4H_4NK. Pyrroles, like phenols, readily couple up with diazomium salts to give azo compounds. In this case the azo – group assumes an α — position, or if both of these are occupied, a β position. Thus pyrrole and benzene diazomium chloride yield pyrrole azo benzene in which the azo groups are in the α - positions. The analogy of phenol is also has been kept in mind, that is on the behaviour of pyrroles towards nitrous and nitric acids; in particular the nitroso pyrroles. Nitroso pyrroles itself is a very unstable compound as can only be obtained by the form of its sodium compound. C_4H_3(:NONa)N.

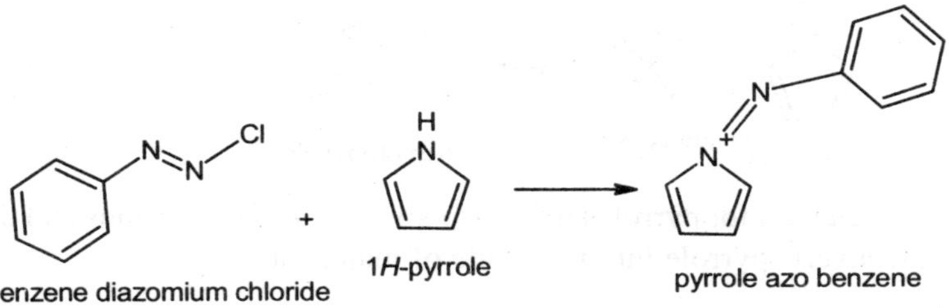

benzene diazomium chloride + 1*H*-pyrrole → pyrrole azo benzene

Pyrrole like phenol and aniline, is readily substituted by halogens all of the four methane hydrogen atoms being replaceable. The most important halogen derivative in tetra – iodopyrrole which is described later in this chapter.

The similarity of pyrrole to aniline is especially evident in its behaviour on alkylation. Organo magnesium halides (Grignard's reagent) react with pyrrole to form magnesium pyrrole compounds of the type.

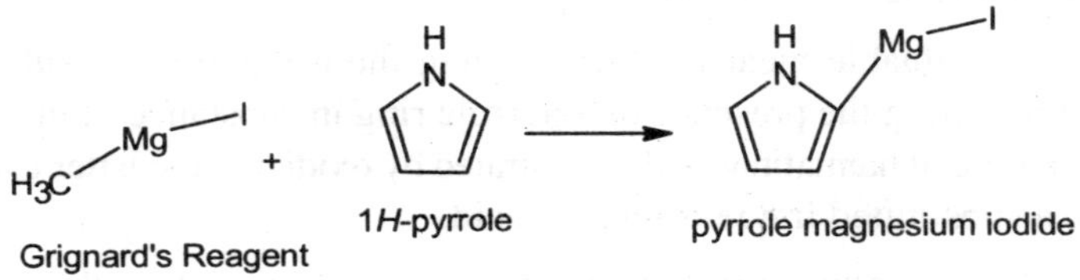

Grignard's Reagent + 1*H*-pyrrole → pyrrole magnesium iodide

These are useful for the synthesis of pyrrole derivatives having side chains in α - position, with carbon dioxide for they give pyrrole - α - carboxylic acid.

Opening of the Pyrrole Ring by Means of Hydroxylamine

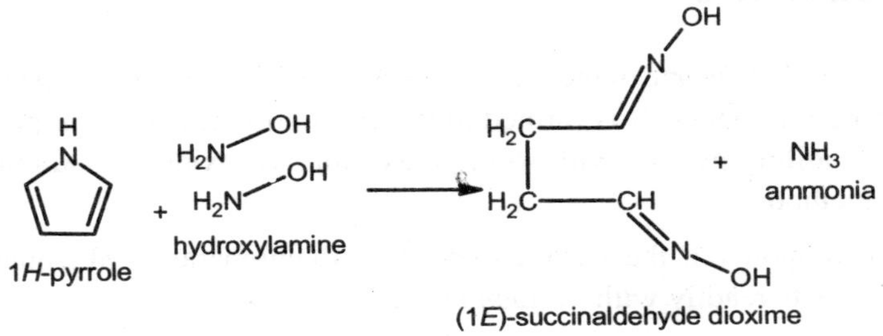

1*H*-pyrrole + hydroxylamine → (1*E*)-succinaldehyde dioxime + NH_3 ammonia

It has been shown by Ciamician that when hydroxylamine reacts with pyrrole the ring is opened and succindialdioxime formed. This reaction is the reverse of the synthesis before and appears to be a general one for pyrrole.

Transformation of the Pyrrole into the Pyrride Ring

When pyrrole or potassium pyrrole is heated with sodium ethoxide and chloroform there is formed β - chloro – pyridine, with methylene iodine, pyridine itself is obtained.

1*H*-pyrrole + chloroform → 3-chloropyridine + Kcl + HCl

This is a general reaction for pyrrole and is also given by its homologous and the indoles. Oxidation with chromic acid converts pyrrole into the imide of maleic acid.

1*H*-pyrrole → 1*H*-pyrrole-2,5-dione

In a similar manner substituted pyrroles, particularly the chloro and bromo – compounds, are oxidized to substituted maleic derivatives.

This oxidation is a valuable means of determining the orientation of substituents in the pyrrole nucleus and also for detecting the presence of a pyrrole ring in substance of unknown constitution. For example the pyrrole nature of hæmatin was demonstrated by oxidizing the latter to a compound $C_8H_9O_4N$, which proved to be a substituted imide of maleic acid.

Tripyrrole $(C_4H_5N)_3$ is produced in certain conditions from pyrrole in the presence of hydrochloric acid. When heated to 300°C tripyrrole decomposes into ammonia, pyrrole and indole.

Other polymers obtained from pyrrole breaks in similar manner, and the process therefore represents *a passage from the pyrrole to the indole series.*

N – Substituted Series

Potassium pyrrole, C_4H_4NK, is one of the most important derivatives of this type and has been repeatedly mentioned in the forgoing pages. It is formed together with hydrogen when potassium is dissolved in pyrrole and also on boiling pyrrole with solid potassium hydroxide. A sodium compound cannot be obtained by these methods.

The potassium compound is the starting material in the preparation of a number of N – derivatives of pyrrole, since it reacts readily with various halogen compounds

$$C_4H_4NK + ICH_3 \rightarrow C_4H_4N.CH_3 + KI$$

N – Alkyl pyrroles are also produced by the above pyrrole synthesis by using alkyl amines in place of ammonia. The N – alkyl pyrroles resemble pyrrole in having feeble basic properties. They do not react with alkyl iodides and are less reactive than pyrrole.

Wandering of Groups from Nitrogen to Carbon

A point of special interest is that N – derivative of pyrrole are transformed under the influence of heat into C – derivative in the same way as alkylated anilines are converted in into C pyridyl – pyrrole

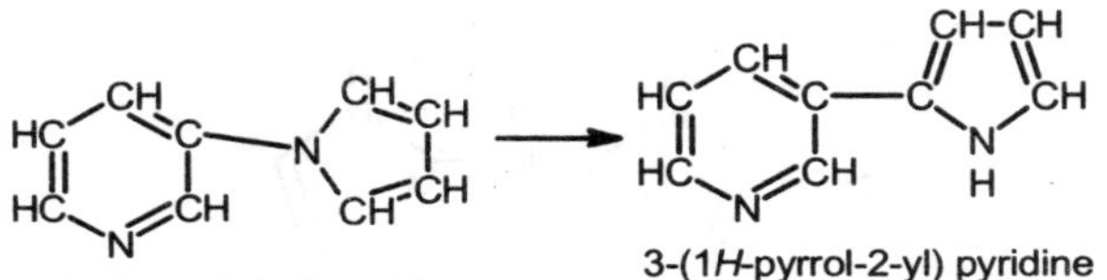

3-(1*H*-pyrrol-1-yl) pyridine → 3-(1*H*-pyrrol-2-yl) pyridine

This conversion fromed one of the stages of Pictets's synthesis of nicotine. The N – Alkyl – pyrroles have lower melting and boiling points than the corresponding C – derivatives.

C – Substituted Pyrroles

Among the halogen derivatives of pyrrole the most important is tetraiodopyrrole or simply **iodole, C_4I_4NH**. It was first prepared by the action of an ethereal solution of iodine on potassium pyrrole (Ciamician and Dennstedt). Later, it was discovered that it could be obtained from pyrrole, potassium hydroxide and iodine and also directly from pyrrole and iodine in the presence of indifferent solvents. It has been employed therapeutically in the treatment of wounds in the place of iodoform, over which it possesses the advantage of being odourless and less poisonous although milder in action. It forms shinning yellowish – brown leaflets which on being heated decompose between 140°C and 150°C, without melting. When reduced with potassium hydrate and zinc dust, tetra – iodopyrrole is converted into pyrrole. With nitric acid or nitrous acid it yields nitro derivatives.

Homologous of Pyrrole

C – Alkyl pyrroles are present with pyrrole in bone oil. Their occurrence thus corresponding exactly to that of toluene and the xylene in coal tar. They may be obtained synthetically by reactions already described in the previous description and are of interest in connection with the structure of hæmoglobin and chlorophyll.

Pyrryl Magnesium Bromide, This compound is produced by the evolution of ethane, by the action of magnesium on pyrrole and with the combination of bromine with the evolved ethylene. This compound can also be used in the preparation of alkyl pyrroles. On interaction with alkyl iodide, for example this compound yields α - allyl and αα' diallyl pyrrole. By treatment with sodium alcoholates it is possible to introduce methyl and ethyl groups into substituted pyrroles.

C – Alkyl pyrrole possesses the same chemical character as pyrrole itself and gives the same colour reactions. In air they change more rapidly than pyrrole, but although resinified with acids are some what more stable in this respect than the parent compound.

Pyrrole derivatives with a methyl group in the α - position and containing at least one unsubstituted nuclear hydrogen atom have been shown to undergo condensation, with aromatic aldehydes however do not attach themselves to the α - methyl group, as is the case with α - substituted pyridine derivatives, nor does the imino group enter into reaction. Two nuclear hydrogen atoms from two pyrrole molecules are eliminated with the aldehydic oxygen atom in the form of water, and the residue unite to give dipyrryl – arryl methane derivatives of the general formula.

In which the coupling may take place in the α – β position of pyrrole.

Pyrrole – 2 – aldehyde

This compound is formed by the interaction of pyrrole and chloroform in the presence of potassium hydroxide solution, thus providing a further illustration of the similarity between pyrrole and the phenols. It crystallizes.

1*H*-pyrrole + chloroform + potassium hydroxide → 1*H*-pyrrole-2-carbaldehyde + 3KCl (potassium chloride) + $2H_2O$ (water)

from cooled petroleum ether in colourless odourless prism like crystals and shows no similarity with benzaldehyde.

Other representatives of this difficulty accessible class has been prepared by H – Fischer; the aldehyde group being introduced into the pyrrole nucleus by Gattermann's Method using anhydrous hydrogen cyanide and dry hydrogen chloride in absolute ethereal solution

Pyrrole Carboxylic Acids

A large proportion of the pyrrole derivatives known at present are carboxylic acid. The simple pyrrole carboxylic acids resemble phenol carboxylic acids are formed by similar reactions e.g.:

1. By the oxidation of homologues of pyrrole by fusion with potassium hydroxide.
2. By treating potassium pyrroles with carbon dioxide

   $C_4H_4NK + CO_2 \rightarrow C_4H_3(COOK)NH$
3. From pyrroles by interaction with carbon tetrachloride and alcoholic potassium hydroxide.

Esters of homologues pyrrole carboxylic acids can be obtained according to the process of pyrrole synthesis described previously when heated, the carboxylic acids rapidly part with carbon dioxide and yield the corresponding pyrrole.

Since pyrrolidine - α - carboxylic acids has been identified as disruption product of proteins and other pyrrolidine carboxylic acids have been discovered in tropionic acid and hygrinic acid, which are degradation products of alkaloids a number of attempts have been made to prepare these acids by hydrogenating the corresponding pyrrole carboxylic acids or esters. So far, however, no satisfactory method of effecting this change has been discovered.

Hydropyrrole Derivatives

Hydropyrrole derivatives (see Pyrrole and Pyrrolidine) are obtained partly by direct hydrogenation of pyrrole derivatives and partly by the degradation of alkaloids, and partly by the synthesis from aliphatic compounds. In addition, certain tetra – hydro – pyrrole bases may be prepared from piperidines which are going to be discussed later, which involve the transformation of a six into five – membered ring. The reduction of pyrrole compounds to di – and especially to tetra – hydro derivatives offers considerable experimental difficulty; in this respect these compounds differ from the pyridine group.

Addition of hydrogen brings about a decided change in chemical nature. Where as pyrrole itself is a quite a weak base, pyrroline and to a still higher degree pyrrolidine possess the strong basic properties of the secondary aliphatic amines. This is the usual consequence of hydrogenating an aromatic system, as it has already been in the case of the naphthalamines and will be observed again in the pyridine group.

The Pyrrolidines

Pyrrolidine and its derivatives have a striking resemblance of piperidines series. This similarity extends even to physical properties and is best illustrated by comparison with the analogy existing between compounds of the pentamethylene and hexamethylene groups.

The discovery of pyrrolidine in 1885 was followed immediately by the recognition of its resemblance to piperidines and its description as a lower nuclear homologue of the latter (Ciamician).

The term nuclear or ring homology, which was originally used only to describe the relationship of the free parent bases, can in the state of our present knowledge be extended compound by compound to the corresponding members of the pyrrolidine and piperidines series. In this manner the similarity is found to hold true for all the more important types of derivatives.

PYRROLIDINE

Introduction

Pyrrolidine is an organic compound with the molecular formula C_4H_9N. It is a cyclic amine with a five-membered ring containing four carbon atoms and one nitrogen atom. It is a clear liquid with an unpleasant ammonia-like odor.

Pyrrolidine is found naturally in the leaves of tobacco and carrot. The pyrrolidine ring structure is present in numerous natural alkaloids such as nicotine and hygrine. It is found in many pharmaceutical drugs such as procyclidine and bepridil. It also forms the basis for the racetam compounds (e.g. piracetam, aniracetam).

A pyrrolidine ring is the central structure of the amino acids proline and hydroxyproline.

Preparation

1. It was first prepared by heating pyrrole with hydroiodic acid and phosphonium iodide at 240° to 250°C. Obviously the less hydrogenated compound pyrroline must be formed as an intermediate product, and this can also be used as starting material.

iodophosphorane + hydrogen iodide + 1H-pyrrole → pyrrolidine + Pyrroline

2. By the reduction of ethylene cyanide, Ladenberg obtained pyrrolidine together with tetramethane diamine.

succinonitrile (ethylene cyanide) + 2H → butane-1,4-diamine (tetramethylene diamine) + pyrrolidine

3. A similar method of preparation was discovered by Gabriel in the interaction of δ - chloro – butylamine (II) and sodium hydroxide.

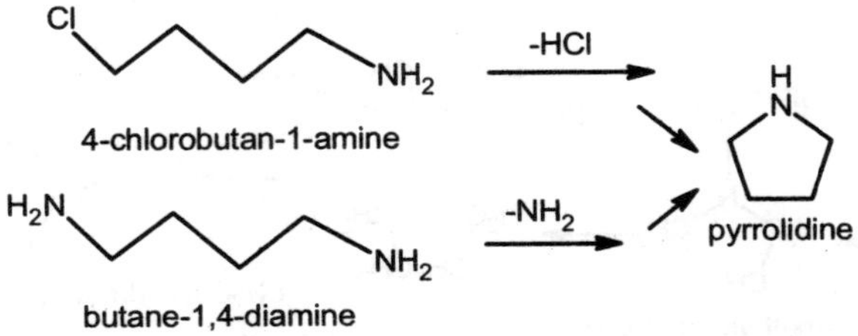

Cl–(CH₂)₄–NH₂ (4-chlorobutan-1-amine) $\xrightarrow{-HCl}$ pyrrolidine

H₂N–(CH₂)₄–NH₂ (butane-1,4-diamine) $\xrightarrow{-NH_2}$ pyrrolidine

Despite the variety of preparative methods available pyrrolidine is difficulty accessible compound. It is a strongly alkaline liquid of boiling point 86°C to 88°C. The physical properties are shown below

Physical Properties

Molecular Formula	= C_4H_9N
Formula Weight	= 71.12096
Composition	= C(67.55%) H(12.75%) N(19.69%)
Boiling Point	= 86 to 88^{0}C
Molar Refractivity	= 21.79 cm^3
Molar Volume	= 84.7 cm^3
Parachor	= 192.9 cm^3
Index of Refraction	= 1.427
Surface Tension	= 26.8 dyne/cm
Density	= 0.839 g/cm^3
Polarizability	= 8.64 cm^3
Monoisotopic Mass	= 71.073499 Da
Nominal Mass	= 71 Da
Average Mass	= 71.121 Da
Log P	= 0.37

Chemical Properties

If dimethyl – pyrrolidium iodide (III) be heated with potassium hydroxide decomposition ensues* resembling that described below under dimethyl piperidinium hydroxide. Under these conditions the ring opens with the formation of an unsaturated aliphatic base, Δ^3 butenyl – dimethylamine (V) (incorrectly called dimethyl pyrrolidine). The methiodide of this base, on distillation with alkali, yields trimethylamine and an unsaturated aliphatic hydrocarbon, known as divinyl (VII). The exhaustive methylation of the pyrrolidine may be represented in the following stages.

---

*results.

I pyrrolidine → II 1-methylpyrrolidine → III cyclopentyl(iodo)dimethylammonium → IV cyclopentyl(hydroxy)dimethylammonium

V *N,N*-dimethylbut-3-en-1-amine → VI 1-hydroxy-1,2,2-trimethyl-1-(1-methylcyclopent-2-en-1-yl)diazanium → VII buta-1,3-diene + propanenitrilium + H_2O

Homologues of Pyrroline

The preparative methods described under pyrroline are also available for formation of its homologues. In addition, a reaction has been discovered by Merling by means of which, the six membered rings can be converted in the five membered ring of pyrrolidine.

Piperidine, or pentamethyleneimine, unites with methyl iodide to form dimethyl – piperidinium iodide, the hydroxide of which (I) on distillation gives an open chain compound (II) known as Δ^4 pentenyl dimethyl – piperidine. The addition product (III) formed by the latter with hydrochloric acid easily isomerizes into methochloride of I: 2 dimethyl – pyrrolidine (IV) which on stronger heating beaks up into methyl chloride and 1:2 dimethyl pyrrolidine.

N,N-dimethylethanamine

chloro(cyclopentyl) dimethylammonium

Just as pyrrolidine on exhaustive methylation gives the unsaturated hydrocarbon divinyl, CH_2:CH.CH:CH2 3 methyl – pyrrolidine gives β - methyl – divinyl CH_2:CH.C(CH_3):CH_2 which is more commonly called *isopirine.*

Pyrrolidine Carboxylic Acids

Certain important dehydration products of the coca and atropa alkaloids have been identified as carboxylic acids of pyrrolidine, viz hygrinic acid and tropionic acid, further it was shown by E. Fischer that pyrrolidine — 2 — carboxylic acid occurs as hydrolysis product of casein, egg albumin, blood fibrin and other albuminous substance when they are treated with hydrochloric acid.

As a result of the above discovered the pyrrolidine carboxylic acids which have previously been little examined. The application of the methods of these carboxylic acids are described below.

PYRROLIDINE-2-CARBOXYLIC ACID

Introduction

It is also called α - pyrrolidine carboxylic acid.

Preparation

Pyrroline was first isolated in an optically impure *l* – form by E. Fischer, from the mixture obtained by hydrolyzing casein with hydrochloric acid. It was since been obtained from a number of other proteins. Hence, it must be assumed that pyrrolidine – 2 – carboxylic acid is a primary product for the hydrolysis of proteins product of the hydrolysis of proteins, and that if is an important unit of the protein molecule.

Na, O, O, CH_3, O, O, CH_3 + Br–CH_2–CH_2–CH_2–Br (1,4-dibromobutane) ⟶ NaBr (sodium bromide) + (2*S*)-pyrrolidine-2-carboxylic acid

Industrial Synthesis of Pyrrolidine – 2 – Carboxylic Acid

I. The synthesis of Willstätter and Ettlinger is an adaptation of the general procedure carried out as follows:

Molecular quantities of trimethene bromide and sodio – malonic ester interest under certain condition to give bromo – propyl malonic ester according to the equation:

This, when treated with bromine, yields $\alpha - \delta$ - dibromo – propyl – malonic ester, which can be considered with ammonia to form the diamide of pyrrolidine of pyrrolidine – 2 – dicarboxylic acid the latter is smoothly converted into pyrrolidine – 2 – carboxylic acid.

2. Fischer has prepared the compound in a similar manner by the interaction of ammonia and phthalimide propyl – bromomalenonic ester.

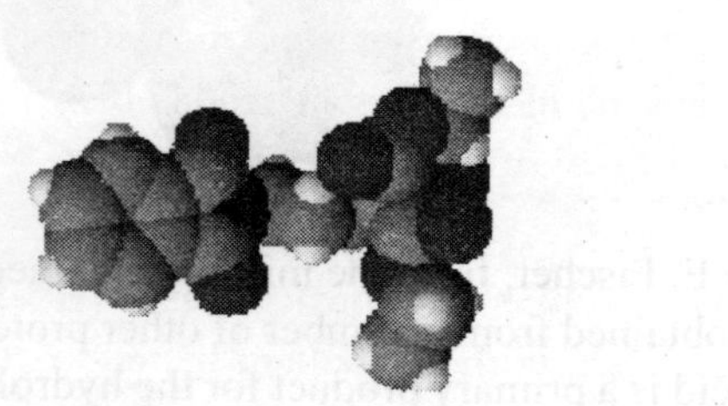

Diethyl bromo [3-(1,3-dioxo-1,3-dihydro-2*H*-isoindol-2-yl)propyl] malonate

3. According to Putochin, proline may be prepared by the direct action of trimethylene bromide on amino – malonic ester.

Naturally these syntheses yield the acid in the racemic form.

Accordingly by carefully drying, *δ, l* pyrrolidine – 2 – carboxylic acid melts with gas evolution at 205°C. In aqueous solution it gives a weekly acid reaction with litmus and possesses a sweet taste. For the separation and identification of the acid the copper salt of service, and has frequently been used by Fischer fir identifying the compound obtained from proteins. If the acid is lead to get saturate with copper hydroxide a dark blue solution is obtained from which the copper salt (dehydrated deposits in deep blue four sided plates. The salt dissolves readily in hot water but sparingly in cold. It is slightly soluble in alcohol and insoluble in chloroform.

Physical Properties

Molecular Formula	= $C_5H_9NO_2$
Formula Weight	= 115.13046
Composition	= C (52.16%) H(7.88%) N(12.17%) O(27.79%)
Melting Point	= 205^{0}C
Molar Refractivity	= 27.90 ± 0.3 cm^3
Molar Volume	= 96.9 ± 3.0 cm^3
Parachor	= 249.0 ± 6.0 cm^3
Index of Refraction	= 1.487 ± 0.02
Surface Tension	= 43.4 ± 3.0 dyne/cm
Density	= 1.186 ± 0.06 g/cm^3
LogP	= –0.57
Polarizability	= 11.06 ± 0.5 $10^{-24}cm^3$
Monoisotopic Mass	= 115.063329 Da
Nominal Mass	= 115 Da
Average Mass	= 115.1305 Da

Chemcial Properties

On being heated for five hours with baryta at 140°C to 145°C, it is converted into the racemic form.

3-chloro-5-(chloromethyl)dihydrofuran-2(3*H*)-one → 4-hydroxypyrrolidine-2-carboxylic acid β' – hydroxy – pyrrolidine - α - carboxylic acid, hydroxy – pyrrolidine has been identified by E. Fischer among the hydrolysis products of gelatin and synthesized by Leuchs in its various stereoisomeric forms by the action of ammonia on α – δ - dichlorovaleroacetone.

TROPIONIC ACID

When tropine and ecgonine are oxidized with chromic acid, they yield dicarboxylic acid compound $C_8H_{13}NO_4$ known as tropionic acids. These differ only in their optical properties the oxidation product from tropine being inactive, and that from ecgonine dexorotatory.

The construction of tropionic acid has been established by Willstätter in the following way :

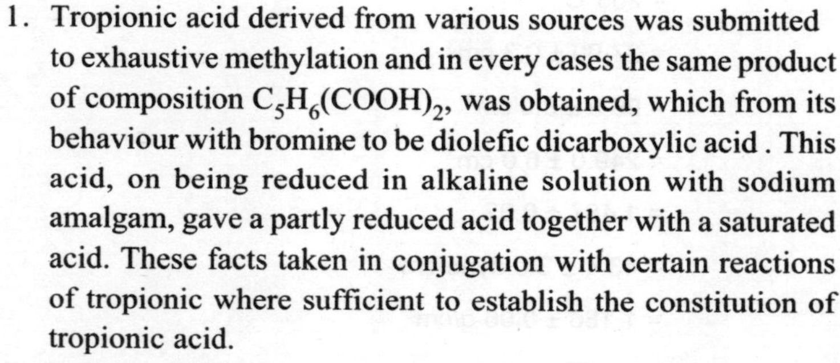

1. Tropionic acid derived from various sources was submitted to exhaustive methylation and in every cases the same product of composition $C_5H_6(COOH)_2$, was obtained, which from its behaviour with bromine to be diolefic dicarboxylic acid . This acid, on being reduced in alkaline solution with sodium amalgam, gave a partly reduced acid together with a saturated acid. These facts taken in conjugation with certain reactions of tropionic where sufficient to establish the constitution of tropionic acid.
2. By treating tropionic acid (or better ecgoninic acid) with chromic acid mixture Willstätter obtained methyl succinimide. The pyrrolidine nucleus of this acid, and therefore of tropine, ecgonine, has thus been isolated in the from of a simple well – known compound.

   Racemic tropionic acid is very much soluble in water but sparingly soluble in alcohol, and insoluble in benzene and ether. It melts indefinitely with decomposition at about 250°C, δ - tropionic acid obtained by the oxidation of either *l* or δ - ecgonine, which melts at 253°C.

Carboxylic Acids of 2 – Ketopyrrolidine (α - Pyrrolidone)

The acids of chef interest in this group are pyrrolidine – 5 – carboxylic acid, a hydrolysis product product of proteins and ecgoninic acid (1 – methyl pyrrolidone – 5 – acetic acid) which an oxidation product of tropine and ecgonine.

Pyrroline – 5 – carboxylic acid, 2 – ketopyrrolidone, 2 – ketopyrrolidine 5 – carboxylic acid. $C_4H_5O(COOH)NH$, was first obtained by heating optically active glutamic acid (α - amino – glutamic acid) at 180°C to 190°C, and was therefore described under the name of pyroglutamic acid. The same acid has also been obtained by heating protein with baryta at 180°C. Subsequently it was prepared from inactive glutamic acid, of which it is the lactams and investigation in greater detail.

2-aminopentanedioic acid → pyrrolidine-2-carboxylic acid + H_2O

ECGONINIC ACID

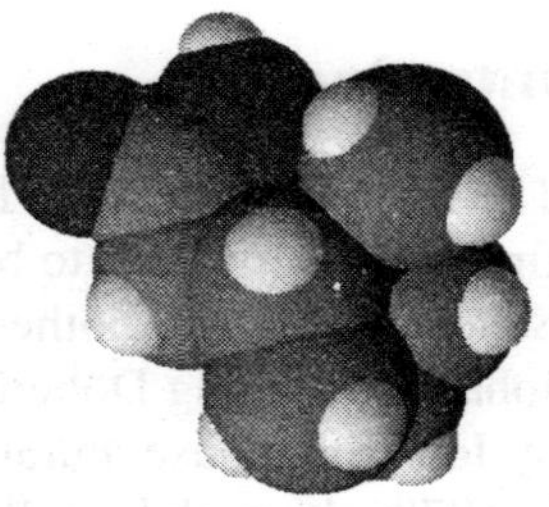

It has already been stated that the oxidation of tropine or ecgonine with chromic acid yields tropionic acid in addition, Liebermann observed that two modifications of ecgoninic acid where produced as by – products in this reaction, l and δ - ecgonines were produces as leavo –rotatory acid. While tropine gave an acid of doubt purity which melts at 90°C. Later it was shown by Willstätter that the ecgoninic acid from the tropine is the racenic form, where as that form ecgonine is the l – variety. The acids are like in all important properties but differ in melting point (r form melts at 93°C) and to some extent in solubility. The above constitution has been confirmed by Willstätter's synthesis of the racemic acid, which was effected in the following manner:

Δ^2 – Dihydro – muconic acid, which has been synthesized from glyoxal and malonic acid, combines with hydrogen bromide to form p – bromo adiphic acid. In methyl alcoholic or benzene solution this readily reacts with methyl amine to give ecgoninic acid probably thorough the intermediate formation of methylamino – adiphic acid.

(I) (2S,3S)-2-bromobutane-1,1,3,4-tetracarboxylic acid + methanamine → 3-(methylamino)hexanedioic acid + methanamine + HBr hydrogen bromide

(II) 3-(methylamino)hexanedioic acid → H_2O + (1-methyl-5-oxopyrrolidin-2-yl)acetic acid

The synthetic product is identical in all respects with that obtained by the oxidation of tropine. This synthesis provided the first direct proof of the existence of a pyrrolidine ring in atropine and cocaine.

Ecgoninic acid crystallizes in white needles. It is very readily soluble in water, alcohol and chloroform, dissolves very sparingly in boiling benzene and is practically insoluble in ether or ligronin. It has no physiological action. The pyrrolidone ring in this compound is so stable that it has not yet been found possible to convert the acid into the open chain β - methylamino – adiphic acid. I — Ecgoninic acid can be obtained as mentioned above, from, *l* or r – ecgoninic acid melts at 117°C.

The alkaloids of the pyrrolidine group are treated in the chapter of alkaloids.

FURANE GROUP

Introduction

The name *furan* comes from the Latin *furfur*, which means bran.* The first furan derivative to be described was 2-furoic acid, by Carl Wilhelm Scheele in 1780. Another important derivative, furfural, was reported by Johann Wolfgang Döbereiner in 1831 and characterized nine years later by John Stenhouse. Furan itself was first prepared by Heinrich Limpricht in 1870, although he called it *tetraphenol*.

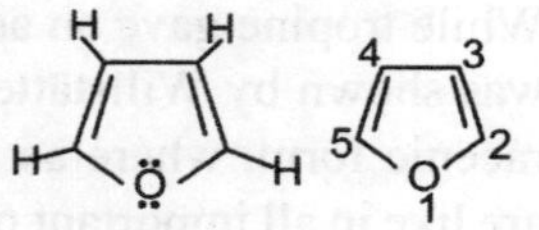

Furan, also known as furane and furfuran, is a heterocyclic organic compound. It is typically derived by the thermal decomposition of pentose-containing materials, cellolosic solids especially pine-wood. Furan is a colourless, flammable, highly volatile liquid with a boiling point close to room temperature. It is toxic and may be carcinogenic. Catalytic hydrogenation of furan with a palladium catalyst gives tetrahydrofuran. Furan is aromatic because one of the lone pairs of electrons on the oxygen atom is delocalized into the ring, creating a 4n+2 aromatic system similar to benzene. Because of the aromaticity, the molecule is flat and lacks discrete double bonds. The other lone pair of electrons of the oxygen atom extends in the plane of the flat ring system. The sp^2 hybridization is to allow one of the lone pairs of oxygen to reside in a p orbital and thus allow it to interact within the pi-system.

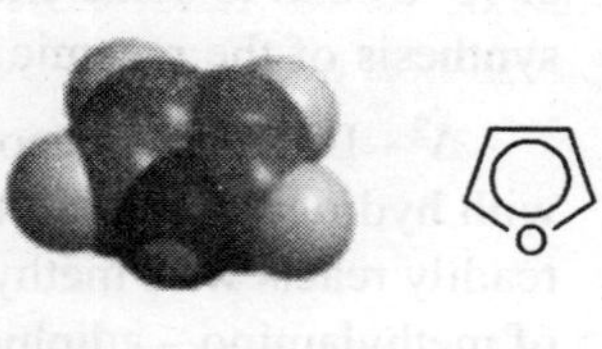

Furane closely relates pyrrole but is of much less importance than pyrrole. It contains a ring composed four carbon atoms and one oxygen atom.

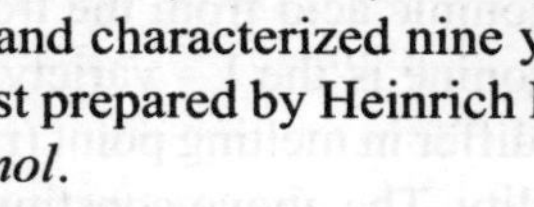

(I) (II)

which may be regarded as a pyrrole ring in which the NH group is replaced by (i)

---

*broken coat of the seeds separated by boiling or other methods (the cotyledons)

The synthetic formation furane derivatives by elimination of water from γ - diketo – compounds has already been discussed in the chapter of aromatic aldehydes and therefore affords a good illustration of the close relationship existing between furane and pyrrole.

Preparation

This is obtained generally from mucic acid. On dry distillation the mucic acid is converted into pyro – mucic acid or furane carboxylic acid, which when heated in a sealed tube yield furane at 270°C

mucic acid → 1-carboxyfuranium Pyromucic acid → furan + CO_2

Physical Properties

Molecular Formula	$= C_4H_4O$
Formula Weight	= 68.07396
Composition	= C(70.57%) H(5.92%) O(23.50%)
Melting Point	= –85.6 °C
Boiling Point	= 31.4 °C
Molar Refractivity	= 18.55 cm^3
Molar Volume	= 72.2 cm^3
Parachor	= 161.2 cm^3
Index of Refraction	= 1.427
Surface Tension	= 24.8 dyne/cm
Density	= 0.942 g/cm^3
Polarizability	= 7.35 cm^3
Monoisotopic Mass	= 68.026215 Da
Nominal Mass	= 68 Da
Average Mass	= 68.074 Da
Log P	= 1.38

Chemical Properties

It is a colourless liquid with a peculiar smell like chloroform. Furane boils at 31.4°C and insoluble in water. The vapour gives a green colouration to the pine stick moistened with hydrochloric acid.

The position of substituents in the furane nucleus is indicated, as in the sane of pyrrole by letters or number (see fig ii)

2,5–DIMETHYLFURAN

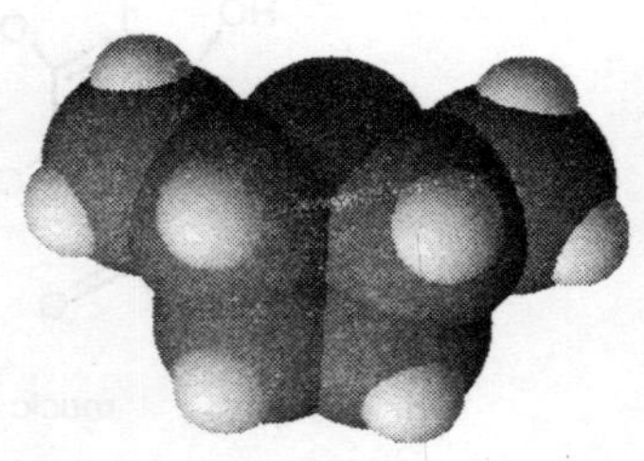

2,5-Dimethylfuran is a heterocyclic compound of the formula C_6H_8O. A derivative of furan, it is also known by its abbreviation **DMF**, not to be confused with dimethylformamide. Recent advances have increased its attractiveness as a biofuel.

Production

Recent research has resulted in improved and more efficient production methods that convert fructose or glucose to a key building block of 2,5-dimethylfuran - hydroxymethylfurfural - in a catalytic biomass-to-liquid process.

Research into an acid catalyst method, undertaken by scientists from the University of Wisconsin-Madison. Another team from Pacific Northwest National Laboratory have developed a non-acidic catalytic method. Although it produces a lower yield, glucose can also be used in the processes.

Potential as a Biofuel

DMF has a number of attractions as a biofuel. It has an energy density 40% greater than ethanol, making it comparable to gasoline (petrol). It is also chemically stable and, being insoluble in water, does not absorb moisture from the atmosphere. Evaporating dimethylfuran during the production process also requires around one third less energy than the evaporation of ethanol,[6][2] although it has a boiling point some 14 °C higher, at 92 °C, compared to 78 °C for ethanol.

The ability to efficiently and rapidly produce dimethylfuran from fructose, found in fruit and some root vegetables, or from glucose, which can be derived from starch and cellulose - all widely available in nature - is likely to add to the attraction of dimethylfuran once safety issues have been examined. Bioethanol and biodiesel are currently the leading liquid biofuels.

Other Uses

2,5-Dimethylfuran acts as a scavenger for singlet oxygen, a property which has been exploited for the determination of singlet oxygen in natural waters. The mechanism involves a Diels-Alder reaction followed by hydrolysis, ultimately leading to diacetylethylene and hydrogen peroxide as products. More recently, furfuryl alcohol has been used for the same purpose.

$^1O_2$ → H_2O → HO … OOH →

2,5 dimethylfuran

diacetylethylene $+H_2O_2$

This compound has also been proposed as an internal standard for NMR spectroscopy. This is due to its good intrinsic properties. 2,5-Dimethylfuran has two singlets in its NMR spectrum at δ 2.2 and 5.8; the singlets give reliable integrations, while the positions of the peaks do not interfere with most analytes. The compound also has an appropriate boiling point of 92 °C which prevents evaporative losses, yet is easily removed.

Toxicology

2,5-Dimethylfuran, together with 2,5-hexanedione and 4,5-dihydroxy-2-hexanone, is one of the main metabolites of hexane in humans, which play a role in the mechanism for the neurotoxicity of hexane.

2,5-Dimethylfuran has been identified as one of the components of cigar smoke with low cilatoxicity (ability to adversely affect the cilia in the respiratory tract that are responsible for removing foreign particles). Its blood concentration can be used as a biomarker for smoking.

TETRAHYDROFURAN

Tetrahydrofuran (THF) (molecular formula: C_4H_8O) is a heterocyclic organic compound. It is a clear, low-viscosity liquid with a smell similar to diethyl ether. It is one of the most polar ethers and is used as a solvent of intermediate polarity in chemical reactions. THF is an aprotic, electron donating solvent with a dielectric constant of 7.6. THF is the fully hydrogenated analog of the aromatic compound furan.

Preparation

THF can be synthesized by catalytic hydrogenation of furan. The major industrial process for making THF is the acid-catalyzed dehydration of 1,4-butanediol. Du Pont developed a process for producing THF by oxidizing n-butane to crude maleic anhydride, followed by catalytic hydrogenation of maleic anhydride to THF.

Precautions

THF tends to form peroxides on storage. As a result, THF should not be distilled to dryness, which can leave a residue of highly-explosive peroxides. Commercial THF is therefore often inhibited with BHT.

FURFURAL

Introduction and History

Furfural was first isolated in 1832 by the German chemist Johann Wolfgang Döbereiner, who formed a very small quantity of it as a byproduct of formic acid synthesis. At the time, formic acid was formed by the distillation of dead ants, and Döbereiner's ant bodies probably contained some plant matter. In 1840, the Scottish chemist John Stenhouse found that the same chemical could be produced by distilling a wide variety of crop materials, including corn, oats, bran, and sawdust, with aqueous sulphuric acid, and he determined that this chemical had an empirical formula of $C_5H_4O_2$. In 1901, the German chemist Carl Harries deduced furfural's structure.

Except for occasional use in perfume, furfural remained a relatively obscure chemical until 1922, when the Quaker Oats Company began mass-producing it from oat hulls. Today, furfural is still produced from agricultural byproducts like sugarcane bagasse and corn cobs.

Preparation

It is formed by the distillation of bran, wood or various carbohydrates with sulphuric, but is best prepared from arbinose or xylenose or from corn cob gum, which is rich in pentoses – treatment with a moderately strong sulphuric acid.

2,3,4,5-tetrahydroxypentanal ⟶ 2-furaldehyde + $3H_2O$

Physical Properties

Molecular Formula	$= C_5H_4O_2$
Formula Weight	= 96.08406
Melting Point	= –36.5 °C
Boiling Point	= 161.7 °C
Flash point	= 162 °C

Composition	= C(62.50%) H(4.20%) O(33.30%)
Molar Refractivity	= 25.30 cm^3
Molar Volume	= 83.8 cm^3
Parachor	= 206.3 cm^3
Index of Refraction	= 1.515
Surface Tension	= 36.5 dyne/cm
Density	= 1.145 g/cm^3
Polarizability	= 10.03cm^3
Monoisotopic Mass	= 96.021129 Da
Nominal Mass	= 96 Da
Average Mass	= 96.0841 Da
LogP	= 0.73

Chemical Properties

In this chemical character furfural is aromatic type and a complete similar to benzaldehyde, Like all aldehydes it forms an oxime and a hydrazine, and in addition undergoes a series of reaction in which the resemblance to benzaldehyde is clearly visible. Thus with alcoholic potassium cyanide it yields furion (C_4H_3O) CHOH.CO (C_4H_3O) an analogue of benzoin

KOH / KCN

2-furaldehyde
furfuran

1,2-di-2-furyl-2-hydroxyethanone
furoin

POISON

With sodium acetate and acetic anhydride it is converted to furfural acrylic acid.

sodium acetate + acetic anhydride + furan → (2Z)-3-(2-furyl)acrylic acid

Uses

The industrial uses of furan are predominantly as an intermediate in the production of tetrahydrofuran, pyrrole, and thiophene; in the formation of lacquers and solvents for resins; in the production of pharmaceuticals; in agricultural chemicals; and in stabilizers (IARC, 1995). Furan is currently produced by one company each in the United States, Belgium, and the Russian Federation. Furan has also been detected in the gas-phase component of cigarette smoke, wood smoke; exhaust gas from diesel and gasoline engines, and in oils obtained by distillation of rosin containing pine wood.

The conversion of pentoses into furfural is employed in determining the proportion of pentoses or pentosans present in vapour mixtures. In this connection a number of insoluble derivatives have been found by means of which furfural may be separated from solution e.g. furfural phloroglucide probably formed according to the equation.

$$C_6H_6O_3 + C_5H_4O_2 \rightarrow C_{11}H_8O_4 + H_2O$$

and the condensation product of furfural and barbituric acid, which is very insoluble.

barbituric acid

1-[(2,4,6-trioxotetrahydropyrimidin-5 (2*H*)-ylidene)methyl]tetrahydrofuranium

Test of Furfuran

A qualitative test for furfuran is the formation of intensely red dye stuff when it is heated with aniline and hydrochloric acid.

PYROMUCIC ACID OR FURAN CARBOXYLIC ACID

It is formed during the dry distillation of mucic acid and also by the oxidation of furfural.

furan $\xrightarrow{CH_3COOH}$ 2-furoic acid (pyromucic acid) + CH_3OH (methanol)

Physical Properties

Molecular Formula	$= C_5H_4O_3$
Formula Weight	= 112.08346
Composition	= C(53.58%) H(3.60%) O(42.82%)
Molar Refractivity	= 25.48 cm^3
Molar Volume	= 84.7 cm^3
Parachor	= 223.3 cm^3
Index of Refraction	= 1.513
Surface Tension	= 48.2 dyne/cm
Density	= 1.322 g/cm^3
Polarizability	= 10.10 cm^3
Monoisotopic Mass	= 112.016044 Da
Nominal Mass	= 112 Da
Average Mass	= 112.0835 Da
Log P	= 0.64

Chemical Properties

From the properties of furfural it might be explained that pyromucic acid would behave as an aromatic acid and as the as the analogue of benzoic acid. This however is not the case. The reactions of pyromucic acid give no indications of aromatic character, but rather place it with the unsaturated aliphatic acids. Thus it immediately decoluorizes the alkaline potassium permanganate solution and when exposed to bromine vapour takes up four atoms of bromine. On warming with bromine water it is converted into fumaric acid on hydrogenation it yields tetra hydro – pyromucic acid.

COUMARONE OR BENZOFURANE SERIES

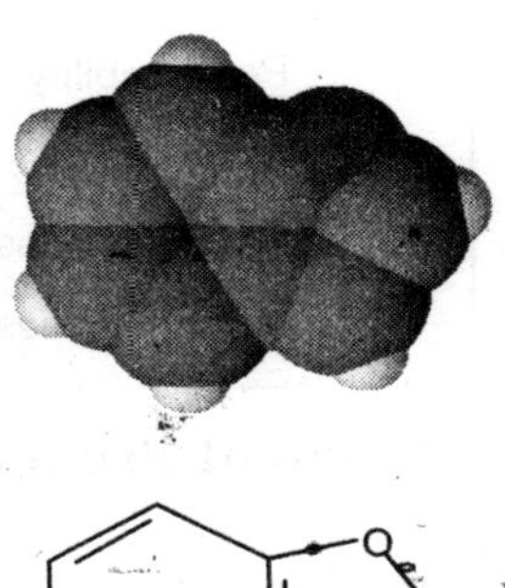

Introduction

Compounds of this type contain a benzene ring and a furan nucleus condensed together with two carbon atoms in common. The parent substance of the group, coumarone thus bears the same relationship to naphthalene as furane to benzene. To furane it is related in the same way as indole to pyrrole.

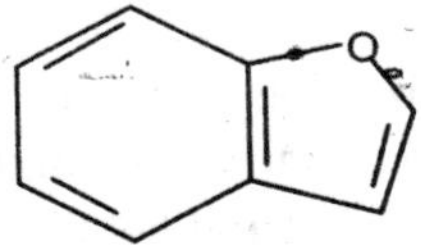

Coumarone derivatives takes their name from their formation by the action of alcoholic potash on α - halogen – substituted coumarins, as a result of which a six membered ring is converted to five membered ring.

Preparation

Coumarone can be prepared by other methods in addition to those described above and is also found together with a number of methyl coumarone in coal tar.

α - Bromocoumarin
KOH
C_2H_5OH
Coumarilic acid
1-benzofuran

Physical Properties

Molecular Formula	= C_8H_6O
Formula Weight	= 118.13264
Melting Point	= < –18 °C
Boiling point	= 173 °C
Composition	= C(81.34%) H(5.12%) O(13.54%)
Molar Refractivity	= 36.39 cm^3
Molar Volume	= 106.3cm^3
Parachor	= 265.0 cm^3
Index of Refraction	= 1.600
Surface Tension	= 38.6 dyne/cm
Density	= 1.110 g/cm^3
LogP	= 2.67
Polarizability	= 14.42 cm^3
Monoisotopic Mass	= 118.041865 Da
Nominal Mass	= 118 Da
Average Mass	= 118.1326 Da

Chemical Properties

It boils at 109°C to 170°C and is extremely stable, indifferent compound. Strong mineral acids bring out resinification and formation of a polymeride known as paracoumarone, through a heated tube when benzene and coumarone is allowed to react, phenathrene is formed.

With chlorine and bromine coumarone yields dihydrogen addition products which on being treated with alcoholic potassium hydroxide, give chloro and bromo – compounds. Nitro – derivatives are also known. A large number of derivatives containing alkyl groups in the benzene and furane nuclei have been prepared by syntheisis.

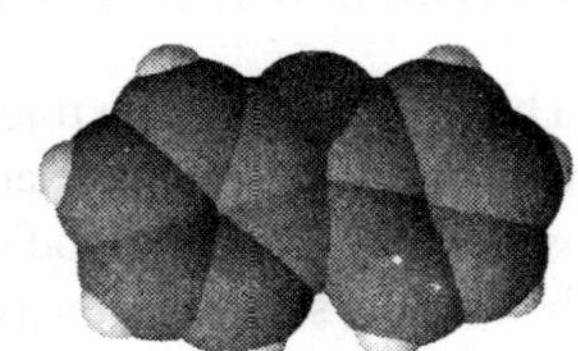

Diphenylene oxide may be regarded as di benzene – furane. It forms white leaflets and is found in coal tar, synthetically, it may be obtained by various methods, from phenol by distillation with lead oxide.

ROTENONE

Introduction

A number of plant products having the properties of fish poisons and insecticides are derived from coumarone. One of the best known of these is rotenone, the active principal of derris root (derris elliptica). The weed grows in the Amazon basin and is used as poison for small fish and for insecticides. It is not poisonous for humans.

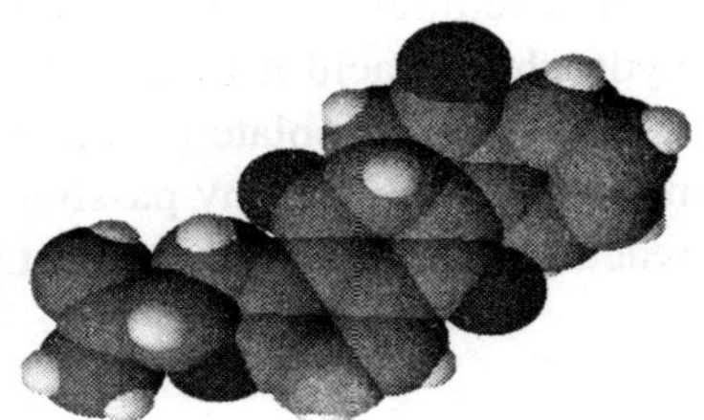

THIOPHENE GROUP

Where as pyrrole as already been said shows resemblance to the phenols, thiophene possesses many points in common with benzene. This similarity extends also to derivatives of these compounds.

It was observed that isatin forms a blue dye if it is mixed with sulphuric acid and crude benzene. The formation of the blue indophenin was long believed to be a reaction with benzene. Victor Meyer in 1883 was able to isolate the substance resposible for this reaction from benzene. This new heterocyclic compound was thiophene

In addition it is found in the lower boiling fractions of the tar from brown coal. Homologous of the thiophene are present in the benzene homologues prepared from coal tar; thus $C_4H_3(CN)_3S$ are contained in commercial toluene and the dimethyl thyophenes or thioxenes $C_4H_2(CH_3)_3S$ in xylol. These are explained by the fact that corresponding homologues of the benzene and thiophene series possess approximately the same boiling points. The thiophene and its homologues may be isolated from the above aromatic hydrocarbons by taking advantage of the fact that they are more readily sulphonated than the latter on treatment with concentrated sulphonic acid.

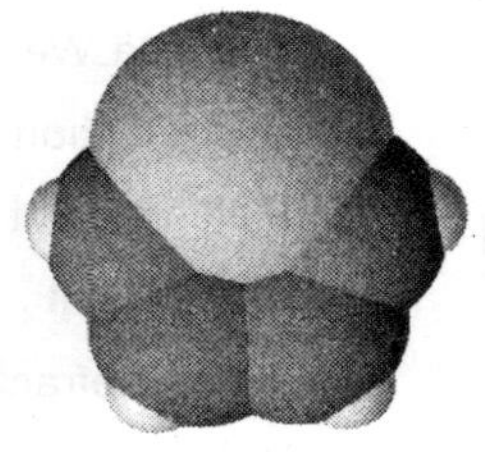

Many bituminous tars and oils such as crude ichtyol oil, consist largely of thiophene homologues. Propyl derivatives of thiophene have been isolated in comparatively large proportion for certain shale tar oils.

Preparation and Extraction

Thiophene can be extracted from commercial benzene by reacted shaking with small quantities of concentrated sulphuric acid, and decomposing the sulphonic acid so obtained by heating it strongly with water. This method of separation is by no means an ideal one, since either a certain proportion to benzene is simultaneously sulphonated, or by using smaller amounts of sulphuric acid, the thiophene is only incompletely removed from benzene.

A better and quantitative method of isolation of thiophene from commercial benzene has been made developed by O. Dimroth. This consists in heating the mixture to the boiling point with a solution of mercuric acetate, when thiophene is attached with the formation of thiophene α – α' dimercuric – hydroxy acetate. The latter separates out as a solid, where as the less reactive benzene is not attacked at this temperature. On heating distilling the mercury compound with moderately concentrated hydrochloric acid it readily decomposes into mercuric chloride and thiophene. This method enables thiophene to be isolated in the pure state without loss of either thiophene vapour or benzene. Thiophene may be synthesized by passing ethyl Sulphide vapour through a red hot tube, and by other pyrogenic reactions. In larger quantities it is obtained by heating sodium succinate with phosphorous trisulphide.

disodium succinate $\xrightarrow{P_2S_3}$ thiophene

Physical Properties

Molecular Formula	= C_4H_4S
Formula Weight	= 84.13956
Composition	= C(57.10%) H(4.79%) S(38.11%)
Melting Point	= –38 °C
Boilig Point	= 84 °C
Molar Refractivity	= 24.63 cm^3
Molar Volume	= 78.9 cm^3
Parachor	= 190.4 cm^3
Index of Refraction	= 1.536
Surface Tension	= 33.9 dyne/cm

Density	= 1.066 g/cm^3
LogP	= 1.90
Polarizability	= 9.76 cm^3
Monoisotopic Mass	= 84.00337 Da
Nominal Mass	= 84 Da
Average Mass	= 84.1396 Da

Chemical Properties

It possesses almost closely resembles in its behaviour towards various reagents. Thus chloro and bromo substitution products can be prepared by the direct action of halogens, in the same way as the corresponding benzene derivative, although the reaction takes place more rapidly than with benzene. Mononitro thiophene melts at 44°C boils at 224°C and has a smell resembling that of nitrobenzene. These compounds however, are not reduced with the sane ease as the nitro benzenes. Since, the majority of reducing agents lead together complete decomposition. The readiness with which thiophene is sulphonated has been mentioned above.

Homologues of Thiophene

These compounds form γ - diketones or alkyl succcinic acid. They may also be obtained from thiophene by reaction similar to those used in preparation of benzene homologues from benzene, e.g. by Fitting Synthesis, from sodium and an alkyl iodide or from thiophene, alkyl bromide and aluminium chloride and so on

1-iodothiophenium + H_3C–I (iodomethane) + 2Na ⟶ 1-methylthiophenium (CH_3–S^+) + 2NaI

Thiophene – aldehyde $C_4H_3(CHO)_3$ can be obtained by the action of hydrogen Sulphide on chlorinated I : 2 – diketo – pentane methylene. In its properties it resembles benzaldehyde rather than furfural.

Thiophene carboxylic acids are prepared in a similar manner to those of the benzene series by oxidizing derivatives of thiophene when the side chain is converted into a carboxylic group. The α acid can also be prepared by the action of sodium on a mixture of iodothiophene and chloro – carbonic ester. It resembles benzoic acid and crystallizes from hot water.

Condensed thiophenes and benzo – thiophenes are also known comparable to the corresponding benzene compounds.

Thionaphthane Thiopthane Diphenylene sulphide

Thionaphthene resembles naphthalene in nature. The true analogue of naphthalene in this series is Thionaphthene, which is formed by heating citric or tricarballylic acid with phosphorous sulphide and is an oil of faint smell. Diphenyl Sulphide, compound of anthracene type is produced when diphenyl sulphide is led through a glowing tube. It melts at 97°C and boils at 332°C.

BENZOPYRROLE OR INDOLE GROUP

The name *indole* is a portmanteau of the words **indigo** and **oleum**, since indole was first isolated by treatment of the indigo dye with oleum.

The molecule of indole contains a benzene nucleus condensed with a pyrrole nucleus as in the formula. As the parent substance of indigo it possess and outstanding interest. Indole and its derivative are also important from the chemo – physiological point of view, owing to their occurrence as disruption products of proteins, normally, indole itself skatole, skatole carboxylic acid and tryptophene. A valuable series of investigations on compounds of indole group was carried out by Bayer in connection with the constitution and synthesis of indigo. The results obtained have contributed not only to our knowledge of indole derivative but also very largely to our process in general organic chemistry. Thus the theory of ring formation gained considerably from the study of indole, and the theory of tautomerism from that of isatin. Further, the results of the researches on indigo lead to the discovery of a number of new methods which have since been applied with success to other branches of organic chemistry.

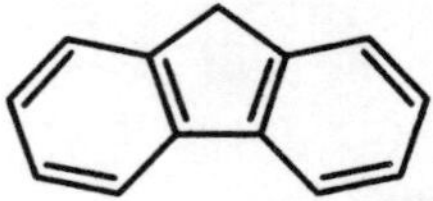

Preparation

It was first prepared by the reduction of indigo blue.

1. From its oxygenated derivative such as oxindole or indoxyl by reduction. For example, by distilling oxindole with zinc dust.

Oxyindole $\xrightarrow[\text{dfistillation}]{\text{Zinc dust}}$ 1*H*-indole

2. By The Leimgruber-Batcho indole synthesis is a series of organic reactions that produce indoles from o-nitrotoluenes **1** The first step is the formation of an enamine **2** using N,N-dimethylformamide dimethyl acetal and pyrrolidine. The desired indole **3** is then formed in a second step by reductive cyclisation.

nitrotoluene —(N,N-dimethylformamide dimethyl acetal / Pyrrolidine)→ 1-[(E)-2-(2-nitrophenyl)vinyl]pyrrolidine —(Raney's Nickel / hydrazine water)→ 1H-indole

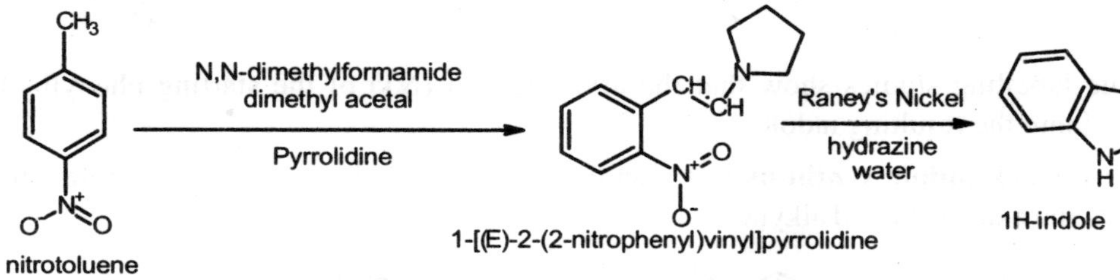

In the above scheme, the reductive cyclisation is affected by Raney nickel and hydrazine. Palladium-on-carbon and hydrogen, stannous chloride, sodium dithionite or iron in acetic acid are also effective reducing agents.

3. By The Fischer indole synthesis is a chemical reaction that produces the aromatic heterocycle indole from a (substituted) phenylhydrazine and an aldehyde or ketone under acidic conditions. The reaction was discovered in 1883 by Hermann Emil Fischer. Today antimigraine drugs of the triptan class are often synthesized by this method.

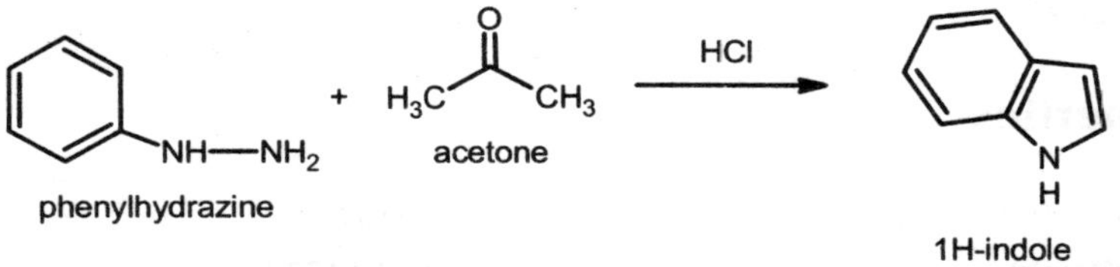

phenylhydrazine + acetone —(HCl)→ 1H-indole

The choice of acid catalyst is very important. Bronsted acids such as HCl, H_2SO_4, and polyphosphoric acid have been used successfully. Lewis acids such as boron trifluoride, zinc chloride, iron chloride, and aluminium chloride are also useful catalysts.

Reaction Mechanism

The reaction of a (substituted) phenylhydrazine with an aldehyde or ketone initially forms a phenylhydrazone which isomerizes to the respective enamine (or 'ene-hydrazine'). After protonation, a cyclic [3,3]-sigmatropic rearrangement occurs producing an imine. The resulting imine forms a cyclic aminoacetal (or *aminal*), which under acid catalysis that eliminates NH_3, resulting in the energetically favourable aromatic indole.

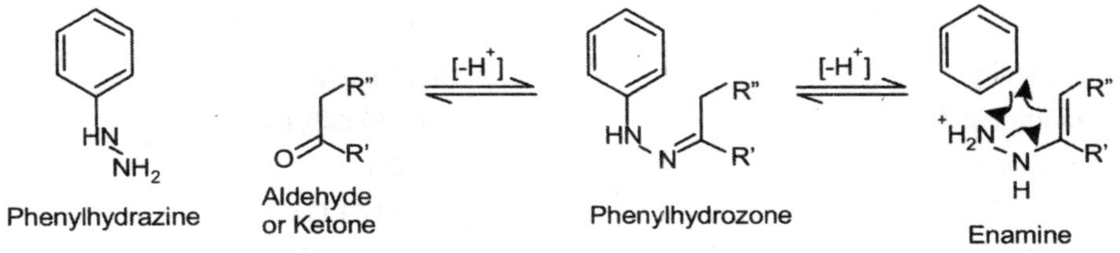

Phenylhydrazine + Aldehyde or Ketone ⇌ $[-H^+]$ Phenylhydrozone ⇌ $[-H^+]$ Enamine

Imine Aminoacetal Indole

Isotopic labelling studies show that the aryl nitrogen (N1) of the starting phenylhydrazine is incorporated into the resulting indole.

4. The **Larock indole synthesis** is a chemical reaction used to synthesize indoles from ortho-iodoanilines and a disubstituted alkyne.

acetylene / lead acetate / base

2-iodo-*N*-methylaniline → 1*H*-indole

An excess of alkyne, using potassium carbonate or potassium acetate as the base, and adding one equivalent of lithium chloride tend to give the best yields. Many functional groups are tolerated on the aniline and the alkyne.

Physical Properties

Molecular Formula	= C_8H_7N
Formula Weight	= 117.14788
Composition	= C(82.02%) H(6.02%) N(11.96%)
Melting Point	= 52 - 54°C
Boiling Point	= 253 - 254°C
Molar Refractivity	= 38.52 ± 0.3 cm^3
Molar Volume	= 101.8 ± 3.0 cm^3
Parachor	= 270.7 ± 4.0 cm^3
Index of Refraction	= 1.680 ± 0.02
Surface Tension	= 49.8 ± 3.0 dyne/cm
Density	= 1.149 ± 0.06 g/cm^3
Log P	= 2.14
Polarizability	= 15.27 ± 0.5 $10^{-24}cm^3$
Monoisotopic Mass	= 117.057849 Da
Nominal Mass	= 117 Da
Average Mass	= 117.1479 Da

Chemical Properties

Indole like pyrrole possesses weakly basic and at the same time somewhat phenolic properties. Similarly, it is easily resinified with acids and gives cherry red colour to a pine splint moistened with hydrochloric acid, substituents being indicated by the use of α , β - N or figures as in the formula. Positions I and 3 are especially reactive, thus with acetic anhydride at 200°C indole yields a mixture of I – acetyl and I:3 diacetyl indole.

1*H*-indole + acetic anhydride → 1-acetyl-1*H*-indole + 1-(1-acetyl-1*H*-indol-3-yl)ethanone 1:3 diacetyl indole

With iodine and very dilute alkali 3 – iodoindole is obtained.

1*H*-indole + I_2 —NaOH→ 3-iodo-1*H*-indole

Electrophilic Substitution

The most reactive position on indole for electrophilic aromatic substitution is C-3, which is 10^{13} times more reactive than benzene. For example, Vilsmeier-Haack formylation of indol will take place at room temperature exclusively at C-3. Since the pyrrollic ring is the most reactive portion of indole, nucleophilic substitution of the carbocyclic (benzene) ring can take place only after N-1, C-2, and C-3 are substituted.

1H-indole —phosphorous oxychloride dimethyl formamide / toluene→ 1H-indole-3-carbaldehyde

Gramine, a useful synthetic intermediate, is produced via a Mannich reaction of indole with dimethylamine and formaldehyde.

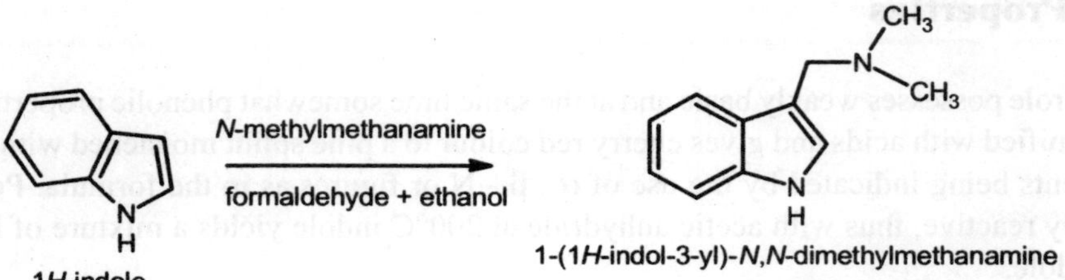

CH_3 N CH_3
N-methylmethanamine
formaldehyde + ethanol
1*H*-indole
1-(1*H*-indol-3-yl)-*N*,*N*-dimethylmethanamine
gramine

The N-H proton has a pK_a of 21 in dimethyl sulphoxide, so that very strong bases like sodium hydride or butyl lithium and water-free conditions are needed for complete deprotonation. Salts of the resulting indole anion can react in two ways. Highly-ionic salts such as the sodium or potassium compounds tend to react with electrophiles at nitrogen-1, whereas the more covalent magnesium compounds (*indole Grignard reagents*) and (especially) zinc complexes tend to react at carbon-3 (see figure below). For the same reason, polar aprotic solvents such as dimethyl formamide and dimethyl sulphoxide tend to favour attack at the nitrogen, whereas nonpolar solvents such as toluene favour C-3 attack.

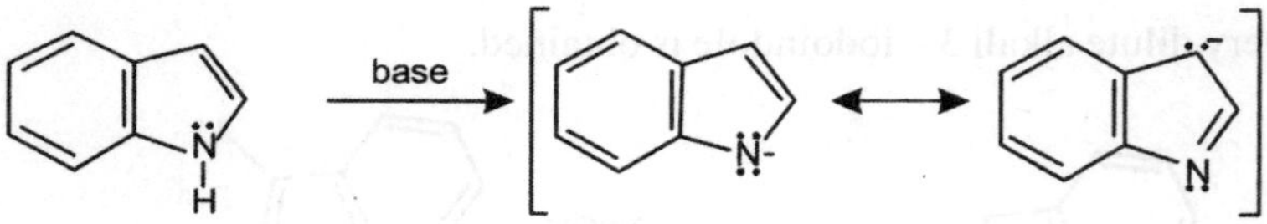

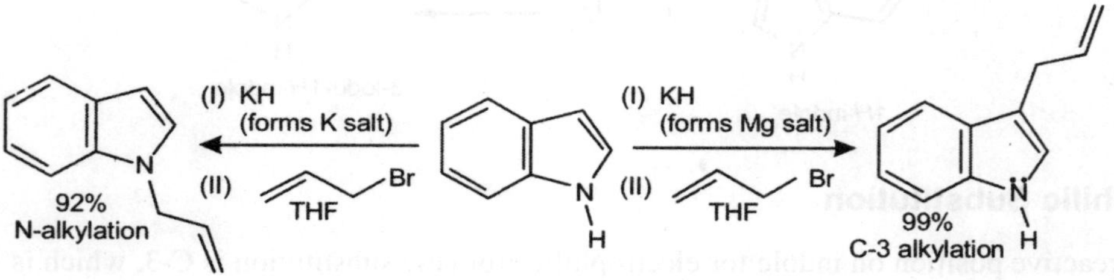

base

(I) KH (forms K salt) (II) Br THF
92% N-alkylation
(I) KH (forms Mg salt) (II) Br THF
99% C-3 alkylation

Carbon Acidity and C-2 Lithiation

After the N-H proton, the hydrogen at C-2 is the next most acidic proton on indole. Reaction of N-protected indoles with butyl lithium or lithium diisopropylamide results in lithiation exclusively at the C-2 position. This strong nucleophile can then be used as such with other electrophiles.

Bergman and Venemalm developed a technique for lithiating the 2-position of unsubstituted indole.[12]

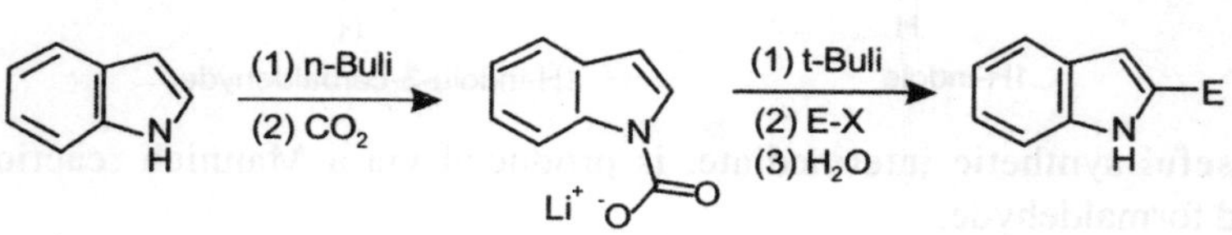

(1) n-Buli (2) CO_2
Li^+
(1) t-Buli (2) E-X (3) H_2O
E

Oxidation of Indole

Due to the electron-rich nature of indole, it is easily oxidized. Simple oxidants such as *N*-bromosuccinimide will selectively oxidize indole **1** to oxindole (**4** and **5**).

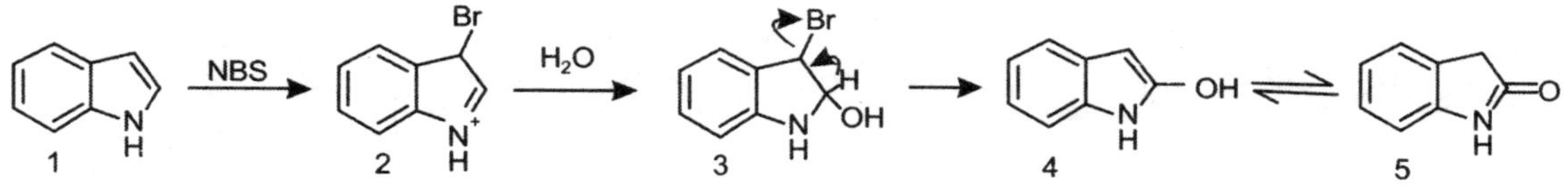

Cycloadditions of Indole

Only the C-2 to C-3 pi-bond of indole is capable of cycloaddition reactions. Intermolecular cycloadditions are not favourable, whereas intramolecular variants are often high-yielding. For example, Padwa have developed this Diels-Alder reaction to form advanced strychnine intermediates. In this case, the 2-aminofuran is the diene, whereas the indole is the dienophile.

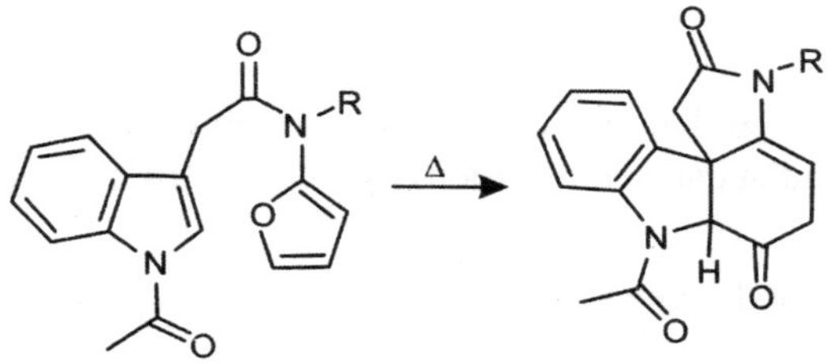

Indoles also undergo intramolecular [2+3] and [2+2] cycloadditions.

Applications Uses

Natural jasmine oil, used in the perfume industry, contains around 2.5% of indole. Since 1 kilogram of the natural oil requires processing several million jasmine blossoms

ALKYL AND ARYL SUBSTITUTION PRODUCTS OF INDOLE HOMOLOGUES OF INDOLE

These generally occur in coal tar and may be prepared from the higher fraction (bp - 250°C - 270°C) of the industrial indole. They are synthesized by the following methods.

1. In an analogous manner to the formation of indole itself by the condensation of aromatic , o – amino compounds.

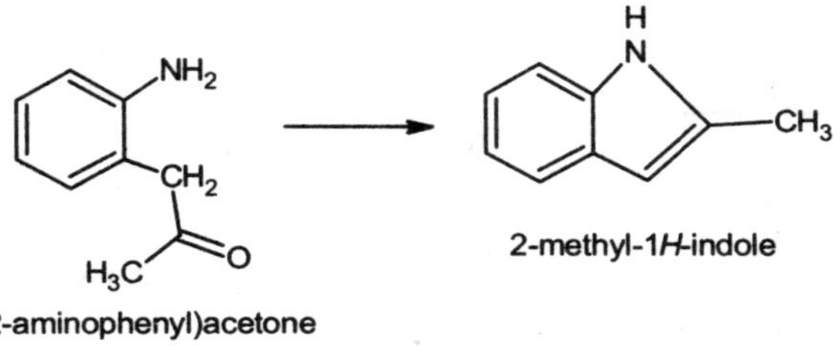

1-(2-aminophenyl)acetone

2-methyl-1*H*-indole

2. By the action of substituted anilines on anilides of ketones or aldehydes of the general formula R'CO.CH(NHC$_6$H$_5$).R". Anilides

aniline → Phenyl anilide → aniline → 4-(anilinomethyl)-N-(phenylmethylidyne) anilinium → α - Phenyl indole

of this type may be prepared from halogen derivatives of ketones and aldehydes, in which halogen is attached to the carbon atom adjacent to the carbonyl group i.e. from compounds containing the group –CO.CH$_3$Cl.

3. By heating the phenylhydrazones of certain aldehydes and ketones of pyrroacemic acid with zinc chloride (Fischer).

1-ethyl-1-methyl-2-phenylhydrazine → 3-methyl-1*H*-indole + NH_3

The true homologues of indole with the substituent in the pyrrole ring are prepared chiefly by methods I and 3. But derivatives with the substituent in the benzene nucleus can also be obtained by these processes, if use is made of starting materials substituted in the phenyl group. Thus 4:7 dimethyl indole has been prepared from the p xylyl – hydrazone of pyroacemic acid. Another very interesting synthesis of this from acetoacetyl acetone and pyrrole.

Acetoacetyl acetone + 1*H*-pyrrole → 2,3-dimethyl-1*H*-indole

Alkyl – indoles resemble the parent compound in being of weakely basic character. The methyl derivatives have a repulsive smell, and on fussion with potash yield indole – carboxylic acids. Most of them give the pine splint reaction.

Indole and its homologues react with chloroform or bromoform in such a eway that the ring is extended and quinoline derivative are produced. Chloroform and α - methyl indole, for example give β - chloro – quinaldine.

3-methyl-1*H*-indole + chloroform → 3-chloro-2-methylquinoline + 2HCl

The distillation of secondary hydro – methyl – indole derivative over zinc dust also leads to the formation of the corresponding quinoline compounds.

2-methylindoline → quinoline

The behaviour of indole and alkyl – indoles on treatment with alkyl iodides is of interest, α, β - Dialkyl – indoles are first formed, with out any alteration of the imnio group e.g. α methyl – indole yields α, β dimethyl indole. This however, is only an intermediate phase. On further interaction with

2-methyl-1H-indole → 2,3-dimethyl-1H-indole → 1-(iodomethyl)-2,3,3-trimethyl-3H-indolium → 1,2,3,3-tetramethyl-3H-1λ^5-indol-1-ol → 1,2,3,3-tetramethylindolin-2-ol → 1,3,3-trimethyl-2-methyleneindoline → 2-isopropylidene-1,3,3-trimethylindoline methyl iodide a rearrangement of the valency bonds takes place (transformation into the pseudoform), the chief product being the methiodide of trimethyl – indoleine, together with a little hydrodide. On treatment with alkalies the first of these undergoes a remarkable change and is converted into trimethyl – methylene – indoleine, which methyl iodide takes up two more methyl groups to form trimethyl – iso – propylidine – iodine. Even this, however, does not represent the end of the reaction since, the last – named compound may on the one hand yield a methiodide, and on the other its hydrodide under the influence of heat may exchange the isopropyl group for once of the methyl group.

SKATOLE

Introduction

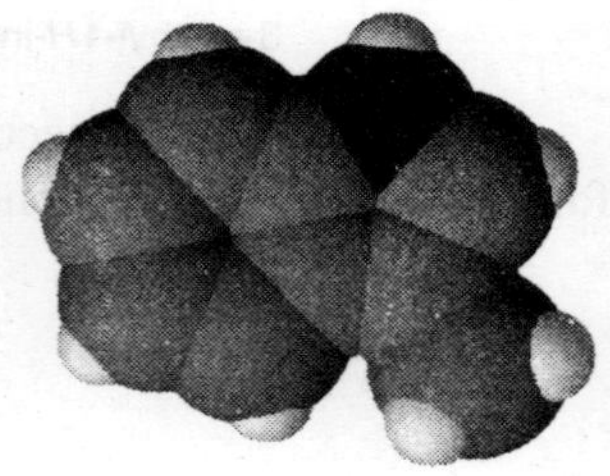

Its name is derived from *scato*, the Latin word for "gush out". As it gushes out in the reaction.

Skatole, β - methyl – indole, is the best known as homologue of indole. It is produced together with a little indole from albuminous matter by fusion with potassium hydroxide and is also present in human fæces. As it can be prepared from propylidine – phenylhydrazones. It crystallizes in plates and has a powerful pungent smell.

Physical Properties

Molecular Formula	= C_9H_9N
Formula Weight	= 131.17446
Composition	= C(82.41%) H(6.92%) N(10.68%)
Melting Point	= 93-95 °C
Boiling point	= 265 °C
Molar Refractivity	= 43.35 ± 0.3 cm^3
Molar Volume	= 118.1 ± 3.0 cm^3
Parachor	= 308.3 ± 4.0 cm^3
Index of Refraction	= 1.654 ± 0.02
Surface Tension	= 46.4 ± 3.0 dyne/cm
Density	= 1.110 ± 0.06 g/cm^3
Log P	= 2.60
Polarizability	= 17.18 ± 0.5 $10^{-24}cm^3$
Monoisotopic Mass	= 131.073499 Da
Nominal Mass	= 131 Da
Average Mass	= 131.1745 Da

Uses

In low concentrations it has a flowery smell and is found in several flowers and essential oils, including those of orange blossoms, jasmine, and Ziziphus mauritiana. It is used as a fragrance and fixative in many perfumes and as an aroma compound.

α Methyl indole: This can be prepared from acetone phynyl hydrazone, the melting point is 60°C and boils at 268°C and resembles indole in smell. N – methyl indole is an oil, boiling point 240°C, formed by eliminating carbon dioxide from the acid obtained from methyl phenyl hydrazine and pyroacemic acid. It has no unplesent odour like skatole.

INDOLE CARBOXYLIC ACIDS

Most of the acids known in this series are prepared by using Fischer's method, which consists in heating the hydrazones of ketonic or aldehydic acids, such as pyroacemic acid and lævailinic acid, with zinc chloride

They may also be prepared in the manner by fusing alkyl indoles with potassium hydroxide, or by the combines action of sodium and carbon dioxide on indoles.

phenyl 2-(phenylhydrazono)propanoate → cyclohexyl 3a,4,5,6,7,7a-hexahydro-1*H*-indole-2-carboxylate + NH_3

methyl (3*Z*)-3-(phenylimino)propanoate → methyl 1*H*-indol-3-ylacetate —Acetic acid→ 1*H*-indol-3-ylacetic acid

The acids are solid, odourless with distinctly acidic and very little basic character. They easily decomposes into carbon dioxide and derivative of indoe.

INDOLE – 3 – ACETIC ACID

Indole-3-acetic acid, also known as IAA, is a member of the group of phytohormones called auxins. IAA is generally considered to be the most important native auxin.

Indole-3-acetic acid, is formed during decomposition of proteins and has been synthesized by Ellinyer according to the methods, just stated above. It crystallizes in small paltes which melts at 164°C and on heating it decomposes into skatole and carbon dioxide.

For along time this acid was assumed by a number of chemists to be indole – 2 methyl 2 – carboxylic acid and it will therefore be found frequently described as skatole carboxylic acid.

Uses and Application

Auxin

It is produced in cells in the apex (bud) and young leaves of a plant. Plant cells synthesize IAA from tryptophan. It has many different effects, as all auxins do, such as inducing cell elongation and cell division with all subsequent results for plant growth and development. There are less expensive and metabolically stable synthetic auxin analogs on the market for use in horticulture, such as indole-3-butyric acid (IBA) and 1-naphthaleneacetic acid (NAA).

Studies of IAA in the 1940s led to the development of the phenoxy herbicides 2,4-dichlorophenoxyacetic acid (2,4-D) and 2,4,5-trichlorophenoxyacetic acid (2,4,5-T). Like IBA and NAA, 2,4-D and 2,4,5-T are metabolically and environmentally more stable analogs of IAA. However, when sprayed on broad-leaf dicot plants, they induce rapid, uncontrolled growth, eventually killing them. First introduced in 1946, these herbicides were in widespread use in agriculture by the middle of the 1950s.

4-CHLOROINDOLE-3-ACETIC ACID

4-Chloroindole-3-acetic acid (4-Cl-IAA) is a natural plant hormone. It is a member of the class of compounds known as auxins and a chlorinated derivative of the more common auxin indole-3-acetic acid (IAA). 4-Cl-IAA is found in the seeds of a variety of plants, particularly legumes such as peas and broad beans.[2][3][4][5] It is hypothosized that 4-Cl-IAA may be a "death hormone" that maturing seeds use to trigger death of the parent plant by mobilizing nutrients to be stored in the seed.

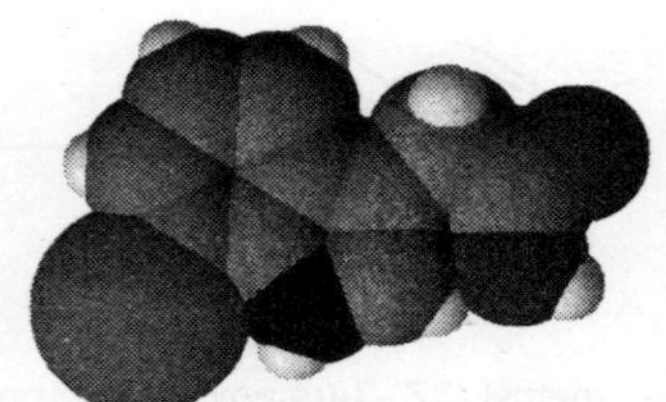

INDOLE-3-BUTYRIC ACID

Indole-3-butyric acid (1H-Indole-3-butanoic acid, IBA) is a white to light-yellow crystalline solid, with the molecular formula $C_{12}H_{13}NO_2$. It melts at 125 °C in atmospheric pressure and decomposes before boiling. Also used as plant hormone

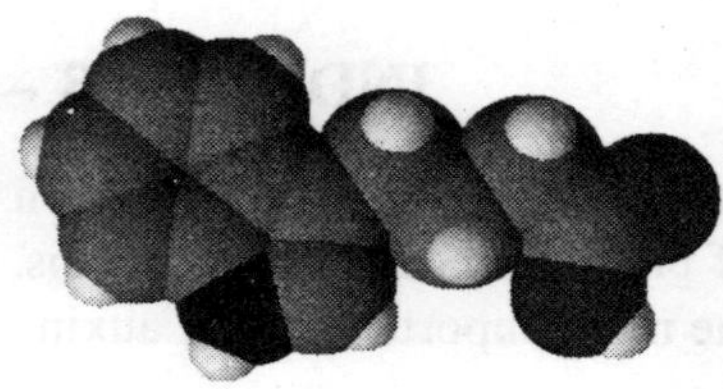

Physical Properties of all the Three above Acids

Properties	4 chloroindole acetic acid	Indole acetic acid	Indole butyric acid
Molecular Formula	$C_{10}H_8ClNO_2$	$C_{10}H_9NO_2$	$C_{12}H_{13}NO_2$
Formula Weight	209.62902	175.18396	203.23712
Melting point		168-170°C	125°C
Molar Refractivity	54.53 cm^3	49.64 cm^3	58.90 cm^3
Molar Volume	141. cm^3	129.3 cm^3	162.3 cm^3
Parachor	404.9 cm^3	369.1 cm^3	448.6 cm^3
Index of Refraction	1.698	1.693	1.645
Surface Tension	67.5 dyne/cm	66.3 dyne/cm	58.3 dyne/cm
Density	1.483 g/cm^3	1.354 g/cm^3	1.252 g/cm^3
Polarizability	21.62 10–24 cm^3	19.67 10^{-24} cm^3	23.35 10^{-24} cm^3
Monoisotopic Mass	209.024356 Da	175.063329 Da	203.094629 Da
Nominal Mass	209 Da	175 Da	203 Da
Average Mass	209.629 Da	175.184 Da	203.2371 Da
Log P	2.53	1.43	2.34

TRYPTOPHANE

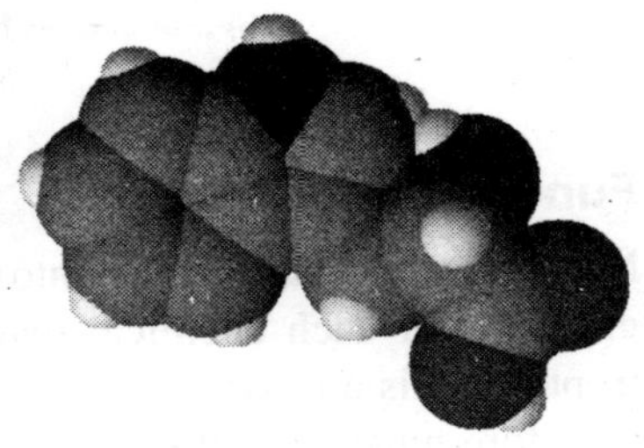

Tryptophane is one of the 20 standard amino acids, as well as an essential amino acid in the human diet. It is encoded in genetic code as the codon *UGG*. Only the L-stereoisomer of tryptophan is used in structural or enzyme proteins, but the D-stereoisomer is occasionally found in naturally produced peptides (for example, the marine venom peptide contryphan). (This is obtained due to the putrefaction* of protein.) The distinguishing structural characteristic of tryptophane is that it contains an indole functional group.

Isolation

The isolation of tryptophane was first reported by Sir Frederick Hopkins in 1901 through hydrolysis of casein. From 600 grams of crude casein one obtains 4-8 grams of tryptophane.

*decay.

Biosynthesis and Industrial Production

Plants and microorganisms commonly synthesize tryptophan from shikimic acid or anthranilate. The latter condenses with phosphoribosylpyrophosphate (PRPP), generating pyrophosphate as a by-product. After ring opening of the ribose moiety and following reductive decarboxylation, indole-3-glycerinephosphate is produced, which in turn is transformed into indole. In the last step, tryptophan synthethase catalyzes the formation of tryptophan from indole and the amino acid, serine.

chlorismate

glutamine glutamine pyruviate

anthranilate

PRPP PP_3

N - 5' phosphoanthranilate

2-{[(1Z,3S,4S)-2,3,4-trihydroxy-4-(phosphonooxy)but-1-en-1-yl]amino}benzoic acid

Glyceraldehyde 3 - phosphate

indole 3 - glyceribe phosphate

1*H*-indole

serine

H_2O

(2*S*)-2-amino-3-(1*H*-indol-6-yl)propanoic acid
triptophane

Function

Metabolism of L-tryptophan into serotonin and melatonin (left) and niacin (right). Transformed functional groups after each chemical reaction are highlighted in red.For many organisms including humans, tryptophan is an essential amino acid. This means that it cannot be synthesized by the organism and therefore must be part of its diet. The principal function of amino acids including tryptophan are as building blocks in protein biosynthesis. In addition, tryptophan functions as a biochemical precursor for the following compounds (see also figure):

- Serotonin (a neurotransmitter), synthesized via tryptophan hydroxylase. Serotonin, in turn, can be converted to melatonin (a neurohormone), via N-acetyltransferase and 5-hydroxyindole-O-methyltransferase activities.

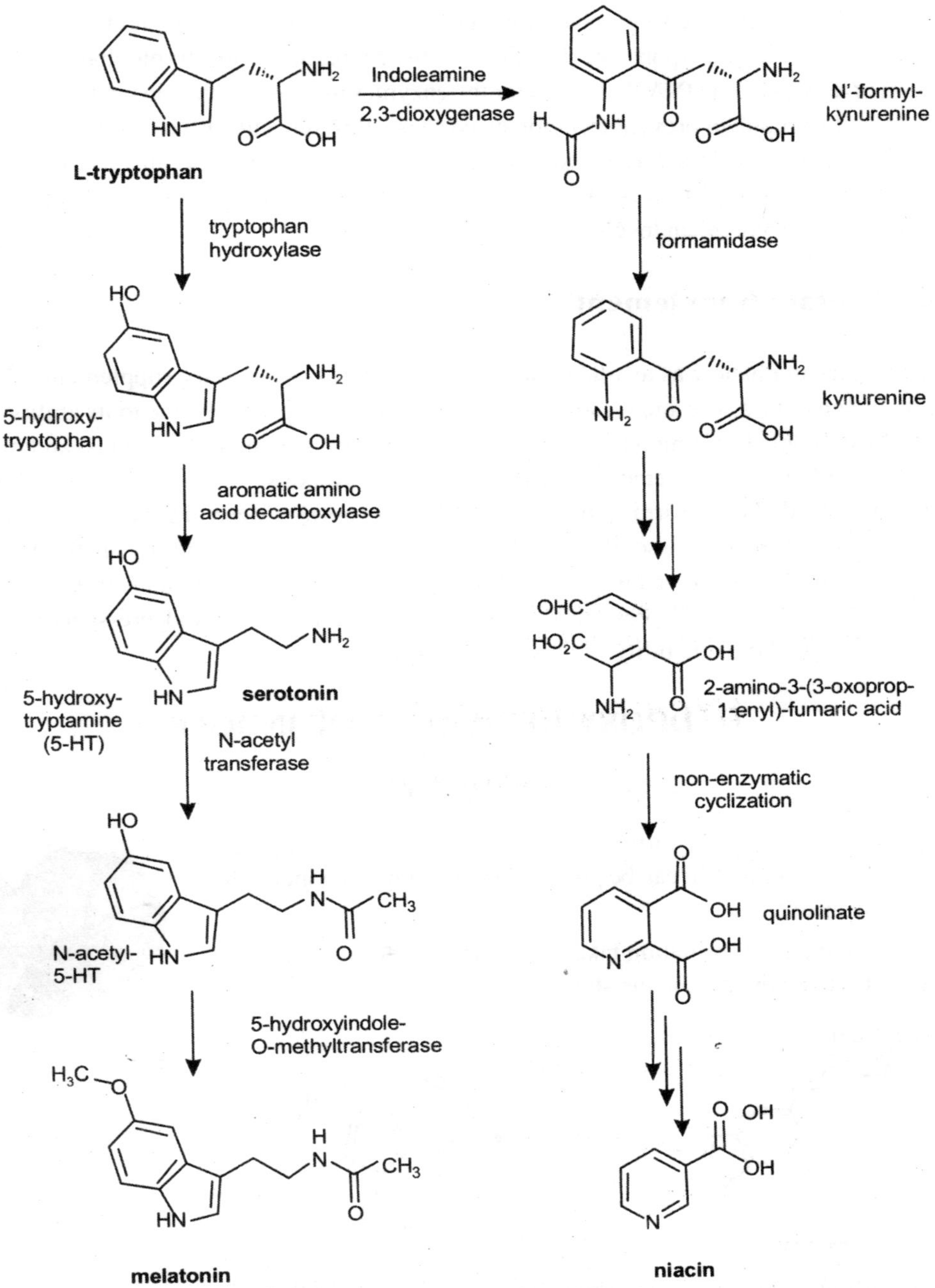

L-tryptophan

Indoleamine 2,3-dioxygenase

N'-formyl-kynurenine

tryptophan hydroxylase

formamidase

5-hydroxy-tryptophan

kynurenine

aromatic amino acid decarboxylase

5-hydroxy-tryptamine (5-HT)

serotonin

2-amino-3-(3-oxoprop-1-enyl)-fumaric acid

N-acetyl transferase

non-enzymatic cyclization

N-acetyl-5-HT

quinolinate

5-hydroxyindole-O-methyltransferase

melatonin

niacin

Niacin is synthesized from tryptophan via kynurenine and quinolinic acids as key biosynthetic intermediates.

In bacteria that synthesize tryptophan, high cellular levels of this amino acid activate a repressor protein, which binds to the trpoperon. Binding of this repressor to the tryptophan operon prevents transcription of downstream DNA that codes for the enzymes involved in the biosynthesis of tryptophan. So high levels of tryptophan prevent tryptophan synthesis through a negative feedback loop and, when the cell's tryptophan levels are reduced, transcription from the trp operon resumes. The genetic organization of the trpoperon thus permits tightly regulated and rapid responses to changes in the cell's internal and external tryptophan levels.

Use as a Dietary Supplement

For some time, tryptophane was available in health food stores as a dietary supplement. Many people found tryptophane to be a safe and reasonably effective sleep aid, probably due to its ability to increase brain levels of serotonin (a calming neurotransmitter when present in moderate levels) and/or melatonin (a sleep-inducing hormone secreted by the pineal gland in response to darkness or low light levels). Clinical research tended to confirm tryptophane's effectiveness as a sleep aid and for a growing variety of other conditions typically associated with low serotonin levels or activity in the brain such as premenstrual dysphoric disorder and seasonal affective disorder. In particular, tryptophane showed considerable promise as an antidepressant alone and as an "augmenter" of antidepressant drugs. However others have questioned the reliability of these clinical trials.

HYDROXY DERIVATIVE OF INDOLE

INDOXYL

This compound is present in the form of potassium salt of indoxyl sulphuric acid, in the urine of mammals it can be prepared by decomposing indoxylic acid with warm water.

Synthetic methods of formation takes place when indicant is allowed to react with dilute acids in aqueous state.

HO OH O HO O OH N H → OH N H

indicane

1*H*-indol-3-ol

It forms bright yellow crystals cannot be satisfactorily distilled, but may be partially volatilized without decomposition by heating in air or with slightly superheated steam. The vapours have fæcal odour.

Physical Properties

Molecular Formula	= C_8H_7NO
Formula Weight	= 133.14728
Composition	= C(72.16%) H(5.30%) N(10.52%) O(12.02%)
Molar Refractivity	= 40.41 cm^3
Molar Volume	= 100.3 cm^3
Parachor	= 285.7 cm^3
Index of Refraction	= 1.739
Surface Tension	= 65.8 dyne/cm
Density	= 1.327 g/cm^3
Polarizability	= 16.01 $10^{-24}cm^3$
Monoisotopic Mass	= 133.052764 Da
Nominal Mass	= 133 Da
Average Mass	= 133.1473 Da
Log P	= 1.41

Chemical Properties

When heated with potassium bisulphite indoxyl yields potassium indoxyl sulphate

potassium bisulphite / warm witrh acid

1H-indol-3-ol ⇌ potassium 1H-indol-3-yl sulfate

which on warming with acids easily reverts to indoxyl. In acid solution indoxyl has strong tendency to resinify and in alkaline medium it is readily oxidized, even by atmospheric air, to give indigo. It is an intermediate in the preparation by the phenylglicine process.

aniline —chloroacetic acid→ phenyl glycene —potassium hydroxide→ 1,2-dihydro-3H-indol-3-one

According to the above formula, which is generally assumed for the solid compound, indoxyl contains the grouping – C(OH) = CH —. The presence of the hydroxyl group can be confirmed experimentally, but in many ways the compound reacts as the keto – form with the group — CO – CH_2 —, Thus giving rise to derivatives of the hypothetical *pseudo – indoxyl.*

INDOXYLIC ACID

Introduction

This is of importance since it occurs as an intermediate product in the technical synthesis of indigo from phenylglicine – o – carboxylic acid. Its ethyl ester is obtained by reducing o – nitrophenyl – propiolic ester with ammonium sulphide

phenyl 3-(2-nitrophenyl)prop-2-ynoate + 4H → phenyl 3-hydroxy-1*H*-indole-2-carboxylate + H_2O

As mentioned above, it readily reacts up into carbon dioxide and indozyl; with oxidsing agent it yields indigo blue, and on heating with sulphuric acid is converted into the sulphonic acid of indigo blue.

OXINDOLE

Introduction

Oxindole is the lactams* or inner anhydride of o – amino – phenyl acetic acid. It is produced by reducing either o – nitro – Phenylacetic acid or dioxindole with tin and hydrochloric, or by isatin with sodium amalgam.

(2-nitrophenyl)acetic acid → (2-aminophenyl)acetic acid → oxindole

*a lactam in which the amide bond is contained within a four-membered ring, which includes the amide nitrogen and the carbonyl carbon. Vividly described in the chapter of chemical reactions and mechanism.

Physical Properties

Molecular Formula	= C_8H_7NO
Formula Weight	= 133.14728
Composition	= C(72.16%) H(5.30%) N(10.52%) O(12.02%)
Molar Refractivity	= 37.29 cm^3
Molar Volume	= 111.0 cm^3
Parachor	= 285.8 cm^3
Index of Refraction	= 1.586
Surface Tension	= 43.9 dyne/cm
Density	= 1.198 g/cm^3
Log P	= 1.15
Polarizability	= 14.78 cm^3
Monoisotopic Mass	= 133.052764 Da
Nominal Mass	= 133 Da
Average Mass	= 133.1473 Da

Chemical Properties

Oxindole possesses a basic as well an acidic character, being soluble in both acids and alkalies. The action of alkalies at high temperature make the indole ring to form salts of o – amino – phenyl acetic acid. Oxidizing agents convert it first into dioxindole.

DIOXINDOLE

It is the lactam of o – amino – mandelic acid and is formed in an analogous manner to oxindole by reduction of a o – nitromandellic acid with zinc dust and acetic acid oxidation it yields isatin, and on reduction oxindole.

o - nitro mandellic acid → o - amino mandellic acid → dioxindole → isatin

ISATIN

Introduction

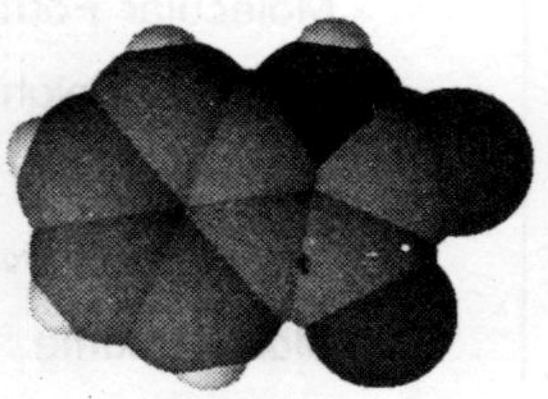

Isatin or **1H-indole-2, 3-dione** is an indole derivative. The compound was first obtained by Erdman[1] and Laurent[2] in 1841 as a product from the oxidation of Indigo dye by nitric acid and chromic acids. The compound is found in many plants and Schiff bases of Isatin are investigated for their pharmaceutical properties.

Preparation

It was observed that isatin forms a blue dye if it is mixed with sulphuric acid and crude benzene. The formation of the blue indophenin was long believed to be are action with benzene. Victor Meyer was able to isolate the substance responsible for this reaction from benzene. This new heterocyclic compound was thiophene.

Isatin is formed by the oxidation of indigotin with nitric acid or chromic acid, and also by oxidation of oxindole and dioxindole. Among synthetic methods of preparation may be mentioned (a) the elimination of water from o – amino benzoyl formic acid.

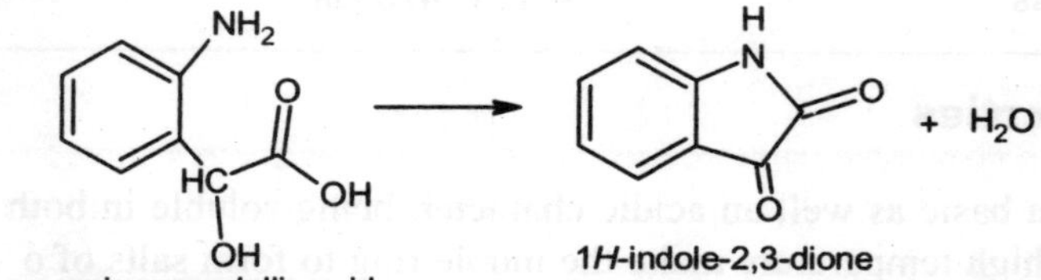

o - amino mandellic acid → 1*H*-indole-2,3-dione + H_2O

and (b) the treatment of o – nitro – phenyl propionic acid with aqueous potassium hydroxide.

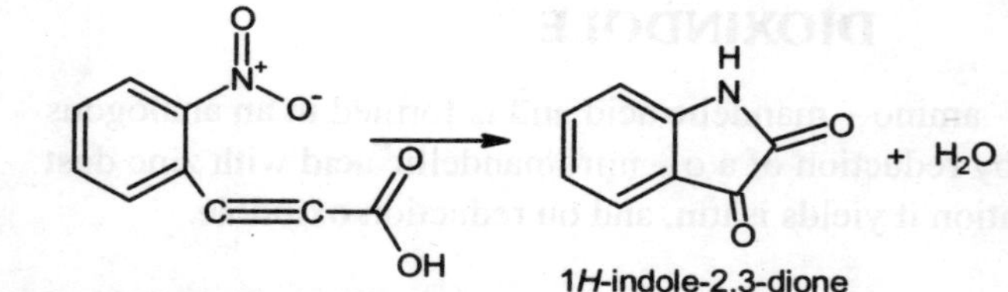

3-(2-nitrophenyl)prop-2-ynoic acid → 1*H*-indole-2,3-dione + H_2O

Physical Properties

Molecular Formula	= $C_8H_5NO_2$
Formula Weight	= 147.1308
Composition	= C(65.31%) H(3.43%) N(9.52%) O(21.75%)

Molar Refractivity	= 37.42 cm^3
Molar Volume	= 107.5 cm^3
Parachor	= 291.0 cm^3
Index of Refraction	= 1.612
Surface Tension	= 53.6 dyne/cm
Density	= 1.367 g/cm^3
Log P	= 0.56
Polarizability	= 14.83 $10^{-24}cm^3$
Monoisotopic Mass	= 147.032028 Da
Nominal Mass	= 147 Da
Average Mass	= 147.1308 Da

Chemical Properties

Isatin dissolves sparingly in water and readily in alcohol and ether, when heated, it volatilizes with partial decomposition. It dissolves in alkalies and if dilute solutions are employed the change from lactam to lactim structure can be followed.

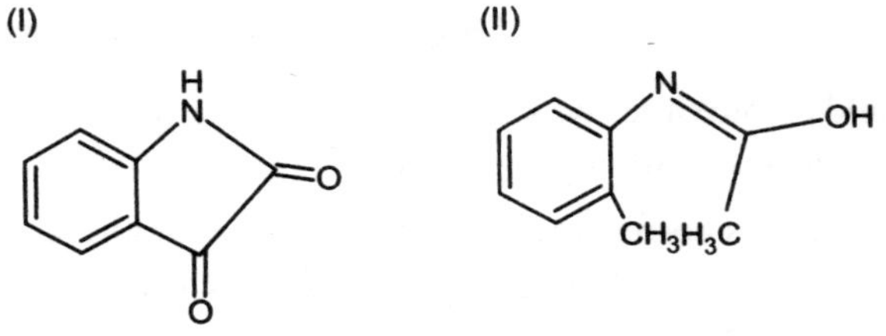

(I) (II)

H N O O N OH CH_3H_3C

The purple red colour, of the N salt passing at the ordinary temperature into the bright yellow of the O – salt. If the solution is warmed the ring opens with the production of the alkali salt of amino – benzoyl – formic acid or isatin. When fused with sodium hydroxide isatin yields aniline,

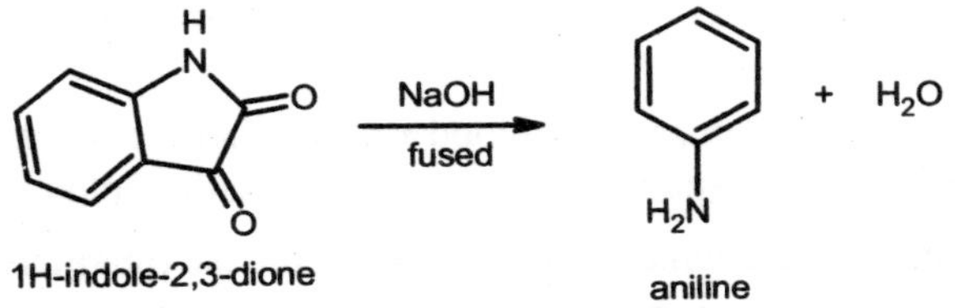

H N O O $\xrightarrow[\text{fused}]{\text{NaOH}}$ H_2N + H_2O

1H-indole-2,3-dione aniline

and with dilute nitric acid it is oxidized to nitro salicylic acid.

1*H*-indole-2,3-dione $\xrightarrow[\text{oxidation}]{HNO_3}$ nitrosalicylic acid

From these reactions this can be decided that the compound contains on nitrogen and one carbon atom in direct union with the benzene nucleus. Bayer's investigation of the reduction productions of isatin – dioxindole, oxindole and indole – finally lead constitution of the compound being established, and there with that of indigo. The above formula for isatin, which had been proposed by Kekule' was also supported by the work of Claisen and Shadwell, who first showed that the isatic acid was identical with o – amino benzoyl – formic acid, and that isatin was its inner aldehyde.

The presence of a keto group in isatin is revealed by the usual reactions. The compound unites with sodium bisulphite and gives a hydrazone with phenyl – hydrazine.

1*H*-indole-2,3-dione + $NaHSO_3$ ⟶ phenylhydrazine + diphenylmethanone hydrazone

with hydroxyl amine it yields isotoxime, which is identical with the nitroso oxinole obtained by the interaction of oxindole with nitrous acid.

A number of isatin derivatives corresponding to the lactin structure are also known. For example when the red silver salt of isatin is treated with methyl iodide the methyl ether of isatin

1*H*-indole-2,3-dione + H_3C–I (methyl iodide) ⟶ 2-methoxy-3*H*-indol-3-one

is formed which crystallized from benzene in blood red prismatic crystals mp 101 - 102°C. When wormed in benzene solution with phosphorous pentachloride, isatin yields isatin chloride

A brown crystalline compound which reverts into isatin on treatment with alkalies, and with ammonium sulphide gives indigo blue, halogens react with isatin to form substitution products and nitric acid converts it into nitro isatin.

1H-indole-2,3-dione + pentachlorophosphorane → 2-methoxy-3H-indol-3-one

Isatin gives a number of colour reactions. It condenses with thiophene to give blue dye stuff indophenine and with pyrrole to yield the blue pyrrole – indophenine

AZOLES

Introduction

Azoles are included in various five membered cyclic systems containing nitrogen. In addition to carbon and nitrogen these rings may also containing oxygen or sulphur, Hence they may be derived from the compounds pyrrole furfurane and thiophene, described in the previous explanations, replacing methane groups with nitrogen atoms. Only the most important of these will be described in detail.

PYRAZOLE GROUP

This group comprises all those compounds, the molecules of which contain a ring composed of three carbon and two nitrogen atoms arranged as follows.

The parent substance of these compounds, pyrazole, is a pyrrole in which a methane group has been replaced by nitrogen.

For this reason the molecular of the pyrrole group is based on that suggested by Knorr for pyrrole derivatives. Just as Dihydro – pyrrole is known as pyrroline and Dihydro pyrazoles are termed as pyrazolines and the completely reduced tetrahydro derivatives, pyrazolidines.

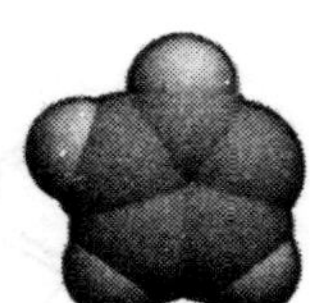

1H-pyrazole

4,5-dihydro-1H-pyrazole
Pyrazoline

pyrazolidine

The position of a substituent group in the pyrazole nucleus is indicated by the numbers 1 to 5 Numbering commences with the nitrogen atom of the imino p group and proceeds in a clockwise direction to the second nitrogen atom, as in the above formula. Ketonic derivatives of pyrazoline and pyrazolidines are usually divided into two classes viz, 4 – derivatives or true ketones, such as keto – pyrazolines and keto – pyralozidine

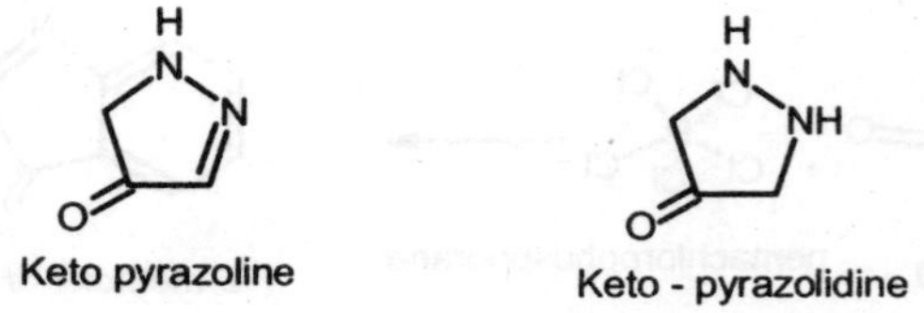

Keto pyrazoline Keto - pyrazolidine

and the 3 and 5 derivative, which are cyclic acid amides for the latter Knorr proposed the names pyrazoline and pyrazolidone.

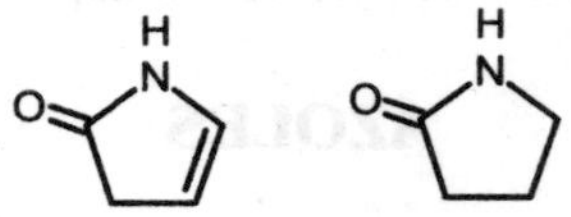

our knowledge of the pyrazole series is large due to the work of Knorr, who described the first representatives of this group in 1883. the physical properties of these compounds and the applications which many of them find their application in medicine among them is antipyrin, for this reason the pyrazole group, after its discovery Knorr was investigated in a number of directions and thanks to the ease with which the pyrazole ring can be formed, these efforts met with considerable success. Among other results, it may be mentioned that Knorr has established the existence of a particular type of isomerism in this series, which appears to throw some light on the structure of benzene.

General Methods of Preparation of Pyrazole and its Derivatives

Various methods of synthesis of pyrazole derivatives have been discovered by Knorr. Esters of β - ketonic acids condense with hydrazines to form derivatives of pyrazolines. The reaction proceeds in two phases. A hydrazone of the ester is first produced, from which alcohol is then eliminated.

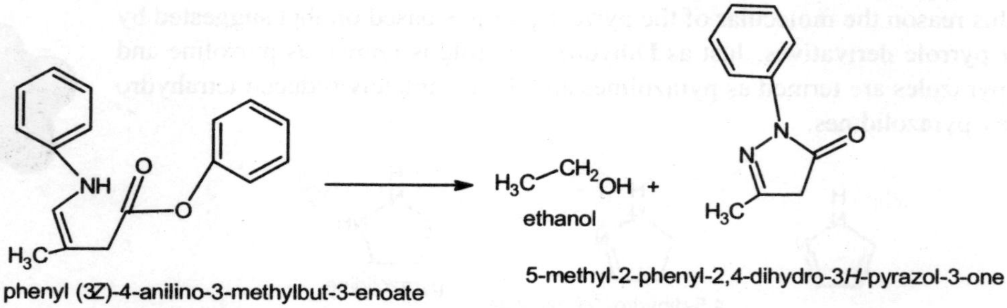

H_3C-CH_2-OH + ethanol

phenyl (3Z)-4-anilino-3-methylbut-3-enoate

5-methyl-2-phenyl-2,4-dihydro-3*H*-pyrazol-3-one

A method similar to the above consists in the interaction of hydrazines with β - keto – compounds of general formula R' – CO.CHR" – COR". This compounds has proved the most fruitful of all reactions devised for the preparation of pyrazole derivatives. On the one hand as basic component, we may employ hydrazine hydrate itself or any primary hydrazine one the other hand, the above general formula includes all the numerous β - ketones and β - keto aldehydes, which can be prepared by the synthetic methods of Claisen and Wislicenus.

3-methylpentane-2,4-dione + ethylhydrazine → 4-ethyl-1-methyl-4,5-dihydro-1*H* -imidazole + $2H_2O$

Hydrazines condense with unsaturated ketones or aldehydes of the type

R' – CO – CR' == CHR" and H – CO – CR' = CHR"

To give derivatives of pyrazole or pyrazoline. Thus parent compound of the later closes pyrazoline, is formed form alcrylein and hydrazine hydrate. The acryl – hydrazine first obtained isomers spontaneously into the cyclic base.

acrylaldehyde + hydrazine → 2*H*-pyrrol-5-amine + 4,5-dihydro-1*H*-pyrazole

Phenyl – hydrazones of unsaturated aldehydes and ketones, containing a double bond in the a - position, may be transformed with great ease into the isomeric pyrazoline derivatives by boiling with glacial acetic acid.

In similar manner unsaturated acids of the acrylic acid series R.CH:CH.COOH react with hydrazines to give derivatives of pyrazoline or pyrazolidones.

2 -butenoic acid + hydrazine → 4,5-dihydro-1*H*-pyrazole pyrazolone

The principle of the above synthesis can be summerised as

Compounds containing two CO group or a Co and a COOH in the b position to one another, or two doubly linked carbon adjacent to a COOH or CO group react with hydrazines to give pyrazole derivatives.

Further synthesis in this group have been affected by E. Büchner in the course of an investigation into the action of diazo – acetic ester on unsaturated compounds. Pyrazole derivatives were obtained.

(*a*) By combination of diazo – acetic acid ester with esters of mono – dibasic acids acetylene series.

dimethyl but-2-ynedioate + methyl 3*H*-diazirene-3-carboxylate → 3-(methoxycarbonyl)-1*H*-pyrazole-4,5-dicarboxylic acid

(*b*) By combining diazo ester of acetic acid with ethylene derivatives such as fumaric acid ester; in this case pyrazoline derivatives are formed.

Fumeric acid ester + Diazoacetic acid ester → pyrazoline - tricarboxylic ester

(*c*) By the action of di azo – acetic ester on ester of certain saturated or unsaturated halogen – substituted acids, α - bromo – cinnamic acid, α – β dibromopropionic acid.

1,3-dibromopropane + Diazoacetic acid ester → pyrazoline - dicarboxylic ester + ethyl bromoacetate + $2N_2$

The preparation of pyrazole itself was first accomplished by the above method E.Büchner obtained it in 1889, by the prolonged action of heat on 3 : 4 : 5 pyrazole – tricarboxylic acid, the ester.

Soon afterwards it was prepared by Balbino by heating hydrazine hydrate with epichlorohydrin and zinc chloride.

Subsequently von Pechmann discovered that acetylene and diazo methane react in a similar manner. This constitutes the simplest synthesis of pyrazole.

acetylene + 3*H*-diazirene → 1*H*-pyrazole

Pyrazole is also obtained by treating the acetyl of proparyl – aldehyde with hydrazine

HC≡C–HC(OCH_2CH_3)$_2$ + Hydrazine → HC≡C—CH=N—NH_2 + 1H-pyrazole

Proparyl acetal; prop-2-ynal hydrazone; 1H-pyrazole

The simple synthesis proves that the atoms in the pyrazole ring are arranged in arrangement with the formula assumed above.

Probably the best means of preparing pyrazole is by heating pyrazole 3 – 5 dicarboxylic acid. In connection of this survey of the synthetic methods available for preparing pyrazole derivative, some reactions of simpler members of the series which are of value in the preparation of simpler members of the series.

A reaction of special importance for the preparation of pyrazole itself consists in the elimination of carboxyl groups from pyrazole – carboxylic acids, by heating the pyrazole – carboxylic acids above the melting points.

The conversion of the oxygen derivatives, pyrazolines and pyrazolidones, into pyrazoles may be affected by distillation with zinc dust or more conveniently by the action of phosphorous pentasulphide or tribromide.

The oxygen of pyrazolines may also be removed by heating these substance with phosphorous oxychloride, when chloro – derivatives are formed.

Physical Properties

Molecular Formula	= $C_3H_4N_2$
Formula Weight	= 68.07726
Melting point	= 66-70 °C
Boiling point	= 168-188 °C
Composition	= C(52.93%) H(5.92%) N(41.15%)
Molar Refractivity	= 18.77 ± 0.3 cm^3
Molar Volume	= 60.9 cm^3
Parachor	= 161.0 cm^3
Index of Refraction	= 1.528
Surface Tension	= 48.6 dyne/cm
Density	= 1.116 g/cm^3
Polarizability	= 7.44 $10^{-24}cm^3$
Monoisotopic Mass	= 68.037448 Da
Nominal Mass	= 68 Da
Average Mass	= 68.0773 Da
Log P	= 0.32

Chemical Properties

The similarity in the formula of pyrazole does not extend to their properties. Pyrazole differ strongly from pyrrole in its remarkable stability and more definitely basic character.

Pyrrole turns brown in air, resinifies with extraodinary ease, and on reduction is converted into di and tetrahydro – derivatives. Pyrazole which crystallizes in long colourless needles is much more resistance to change.

In pyrazole the basic character is very less. On the other hand pyrazole, although it gives no reaction with litmus and can be removed weakly acid solution by a current of steam, nevertheless it yields well defined salt and acids.

All the chemical properties of pyrazole show it to be more nearly allied to pyridine and benzene than to pyrrole.

It exhibits to an even greater degree than thiophene, those peculiarities which were first observed in the aromatic series and are therefore associated with the term "aromatic character".

A number of facts established by Knorr, clearly illustrate the aromatic character of pyrazole.

1. Fuming sulphuric acid converts pyrazole into a sulphonic acid which in its reactions shows certain resemblances to the aromatic sulphonic acids.
2. In halogen derivatives of pyrazole a halogen atom attached to the nucleus is even more firmly held than in the benzene derivatives.
3. When pyrazole is treated with con nitric acid, hydrogen is readily exchanged for a nitro – group. Like the aromatic nitro compounds, 4 – nitro pyrazole and its derivatives can be reduced to amino compounds.
4. Amino pyrazole resembles the aromatic bases in its behaviour, it gives a colour reaction with a solution of bleaching powder and its readily diazotized.
5. Diazo – pyrazoles can be coupled with phenols to form azo – dye in exactly the same manner as the aromatic diazo compounds. They differ from most of the latter in the stability their salts in aqueous solutions. On boiling these solutions there is no visible evolution of nitrogen; these occur on prolonged heating at a high temperature. Diazo pyrazoles however, do not under go the usual " diazo reactions".
6. Pyrazolone or 5 – Hydroxy – pyrazole, has a pronounced phenolic character

The similarity between pyrazole and pyridine may also be illustrated by several examples. It is clearly visible from the behaviour of the double salts formed by pyrazole with platonic chloride, an in similar double salts given by pyridine and pyrazole with other metallic compounds, such as mercuric chloride, potassium platinous chloride and the sulphate of copper, zinc and cadmium.

In smell and other properties the alkyl derivatives of pyrazole closely resemble pyridine bases that on casual examination they may easily be mistaken for them. The carboxylic acids of pyrazole also possesses many points in common with those of pyridine, for example, when polycarboxylic acids are heated, they part with carbon dioxide to give the mono acids, and is found that the carboxylic group the a - position to nitrogen is the first to be removed, By a repetition of the process the mono- carboxylic acid is converted into pyridine. The properties of 3, 5 or 5 pyrazole – carboxylic acid are closely allied to those of pyridine α - carboxylic acid.

Owing to the basic character of pyrazole, it resemblance to pyridine is more evident than its resemblance to benzene. Pyrazole is a weak secondary base; as such it may be acylated, benzylated and then converted into derivatives of urea and urethane.

It unifies with alkyl iodides to form crystalline ammonium compounds. These are of importance in the preparation of homologues of pyrazole, as under the influence of heat the acetyl radical is transferred from nitrogen to a carbonation of the nucleus. The Hoffmann synthesis of aniline homologues can therefore be applied to pyrazole series. As will be seen later, this reaction is also of value in the pyridine group.

A separation of the secondary and tertiary pyrazole – bases resulting from the above process may be effected by taking advantage of the fact that secondary pyrazoles can be quantitatively thrown out of an aqueous solution in the form of their silver salts, tertiary pyrazoles remaining unchanged.

The silver compounds are readily formed by all pyrazoles containing a free imino – hydrogen atom, and are useful for the preparation of N – alkyl substituted derivatives by double decomposition with alkyl iodides. Thus silver pyrazole and methyl iodide yield 1 – methyl pyrazoles.

Ag–pyrazole + H_3C–I (iodomethane) ⟶ 1-methyl-1*H*-pyrrole (CH_3–N ring) + Ag–I

According to Knorr, these N – Alkyl are better prepared by distilling the corresponding pyrazole alkaloids.

An interesting regularity has been observed in connection with the physical constants of pyrazole homologues. Symmetrically constituted compounds possess higher melting points than the isomeric unsymmetrical compounds and of the latter the tertiary derivatives melts much lower than the isomeric secondary bases

A further difference between c – alkyl and N – alkyl derivatives is that the former usually have only a slight odour, while the latter generally have a strong smell recalling that of pyridine.

Pyrazoles	Melting point	Boling Point
3:5 Dimethyl – pyrazole (sym)	107	220
3:4 Dimethyl – pyrazole (unsym sec)	57 to 59	222
1:3 Dimethyl – pyrazole (unsym tert)	liquid	140°C

It has already been mentioned that the pyrazole nucleus is stable towards oxidizing agents. The alkyl groups attached to the ring in homologues of pyrazole may be successfully oxidized to carbonyl group by use of potassium permanganate.

In 1 – phenyl – pyrazoles the benzene nucleus is more readily oxidized away than the pyrazole nucleus, as is shown by the formation of pyrazole when 1 – phenyl – pyrazole is oxidized with potassium permanganate in sulphuric acid solution (Knorr) As in the aromatic series, the benzene ring is still more easily disrupted if it is stability is first lowered by the introduction of an amino or hydroxyl group. For example, the benzene ring of 1 – amino – phenyl 3 – methyl pyrazole is much more readily attached than that of 1 – phenyl – 3 – methyl – pyrazole. A remarkable point is the stability of pyrazole derivatives under these conditions as compared with corresponding derivatives of pyrazole, the latter being completely oxidized by potassium permanganate.

The action of nascent hydrogen (form sodium and alcohol) varies with different compounds of pyrazole group. Pyrazole itself and its homologues are apparently not attacked by sodium and alcohol. 1 – phenyl pyrazole and its homologues, i.e. those derivatives formed by the interaction of phenyl – hydrazine and β - diketone compounds have been shown by Knorr to be reduced to pyrazoline derivatives.

1-phenyl-1*H*-pyrazole + 2H → 1-phenyl-4,5-dihydro-1*H*-pyrazole

Pyrazoline bases obtained from phenyl hydrazine are changed by oxidizing agents, such as chromic acid nitrous acid, ferric chloride and hydrogen peroxide, into characteristic dyes varying from red to blue in colour. This reaction described by Knorr as the **Pyrazoline reaction**, this may be used for the detection of pyrazole and pyrazoline bases derived from phenyl – hydrazine.

The reaction is conveniently carried out as follows: A small amount of pyrazole base is dissolved in alcohol in a test tube, and the mixture is boiled and a small piece of sodium added to the boiling mixture, after cooling the solution is diluted with water and the alcohol is distilled off, and the pyrazole base is extracted by the means of ether, the base after removing is dissolved in comparatively strong sulphuric acid and a drop of a solution of sodium nitrite or potassium dichromate added produces a fine red or blue colour.

Pyrazole and its Derivative

Pyrazoline derivatives differ considerably in their properties from those of pyrazole, owing to their much lower stability.

This is another indication of the aromatic nature of the pyrazole, since it is almost a characteristic of aromatic compounds that the addition of two hydrogen atoms to the ring results in diminished stability.

The pyrazolines give the reaction of aliphatic derivatives, resembling unsaturated compounds in their behaviour towards potassium permanganate and nascent hydrogen. They resemble hydrogen in the manner in which they are hydrolyzed by mineral acids, and aldazines in their decomposition into gaseous nitrogen and nitrogen free substance. The presence of a five – membered ring is only revealed in the ease with which pyrazoles are converted into pyrazoles.

Pyrazoline and its homologues are weak bases. In general they only dissolve in concentrated acids, forming unstable salts which dissolve on the addition of water. The parent substance, pyrazoline, boiling point 144°C is the most stable of all these compounds.

The pyrazolidines or completely reduced pyrazoles have not been thoroughly investigated owing to their instability. They possess strong reducing properties and readily give up hydrogen to form pyrazolidines.

Derivatives of Pyrazoles – Tautomerism in Pyrazole

As the result of an investigation into the synthetic derivatives of 1 – phenyl pyrazole. Knorr assigned to the then unknown pyrazole the following constitution. The problem of the structure of pyrazole entered a new phase in 1893 when Knorr and MacDonald showed that the oxidation of the well known isothermic compounds 1 – phenyl – 3 – methyl – pyrazole, or their amino – derivative, give one and the same methyl pyrazole of boiling point 204°C

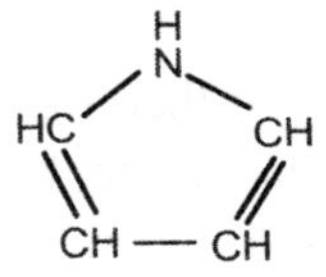

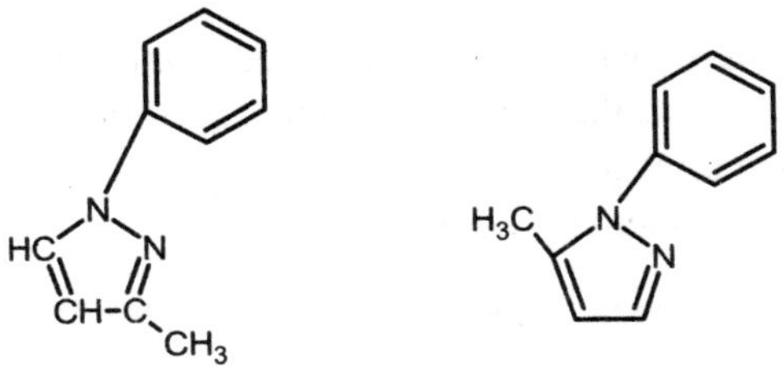

3-methyl-1-phenyl-1*H*-pyrazole

5-methyl-1-phenyl-1*H*-pyrazole

By this method it is therefore not possible to obtain the two methyl pyrazoles of the formula corresponding to the above phenyl methyl pyrazoles.

Knorr and MacDonald also failed to observe the production of isomeric methyl – pyrazoles on condensing hydrazine hydrate with oxy methylene acetone. On the other hand, Claisen and Roosen found that phenyl – hydrazine reacted with oxy – methylene acetone to form two isomeric phenyl methyl pyrazole.

3-methyl-1*H*-pyrazole and 5-methyl-1*H*-pyrazole

Methyl – pyrazole may therefore be a mixture of the two desmotropic* forms 3 methyl pyrazole and 5 methyl pyrazole, in a state of continuous and rapid interconversion.

For these reason Knorr described the compound as 3 – (5) methyl pyrazole and formulated it as I below:

I: H, N, N, CH, C. CH_3, CH

II: H, N, N, CH, CH, CH

The accuracy of this conclusion was confirmed by the discovery that methyl – pyrazole could react simultaneously in the sense of a 3 and 5 methyl pyrazole.

It is therefore assumed that the 1 – hydrogen atom in pyrazole is not permanently attached to given nitrogen atom, but is linked sometimes to the other, with necessary readjustment of the double bonds, as indicated in formula II. But the actual position of equilibrium varies with the nature of the substituent group or groups attached to the pyrazole ring and may in extreme cases correspond to a definite fixed structure.

In addition to the above tautomerism has been observed in the pyrazole series in connection with 1 – phenyl – 3 – methyl 5 – pyrazolone the parent compound of antipyrine, which is usually formulated as 1, it is prepared in large quantities as in intermediate product in the manufacture of antipyrine, by condensing acetoacetic ester with phenyl – hydroazine.

According to Knorr this extremely reactive compounds may react simultaneously in the three tautomeric forms.

In this case we are therefore we are dealing with a very complicated case of tautomerism, described by Knorr as "double tautomesim"

I – Phenyl – 3 – methyl – 5 – pyrazolone itself is known only in one form whether prepared by synthesis or form a derivative corresponding to one or other of the above three types, it is always obtained in the form of a substance of melting point 127°C, crystallizing in white prisms which of

*Tautomerism in which both tautomeric forms have been isolated.

above three formula represents this compounds has not yet been established with certainly, but numerous derivatives are known corresponding to each type.

Methyl form

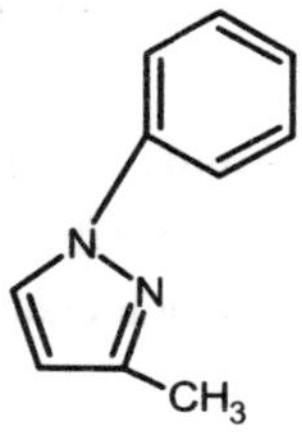

Imine form

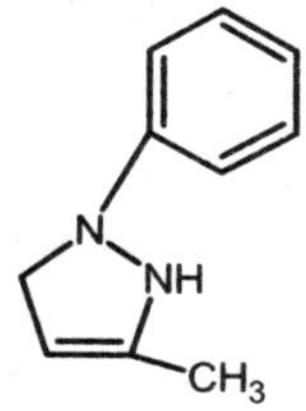

Phenolic form

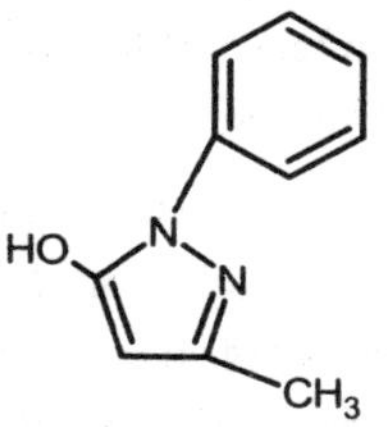

3-methyl-1-phenyl-1*H*-pyrazol-5-ol

5-methyl-2-phenyl-2,3-dihydro-1*H*-pyrazole

The majority of the derivatives of phenyl – methyl pyrazolone and pyrazole, blue. The latter is readily obtained by gentle oxidation of phenyl – methyl pyrazolone, and represents the indigo of the pyrazole series.

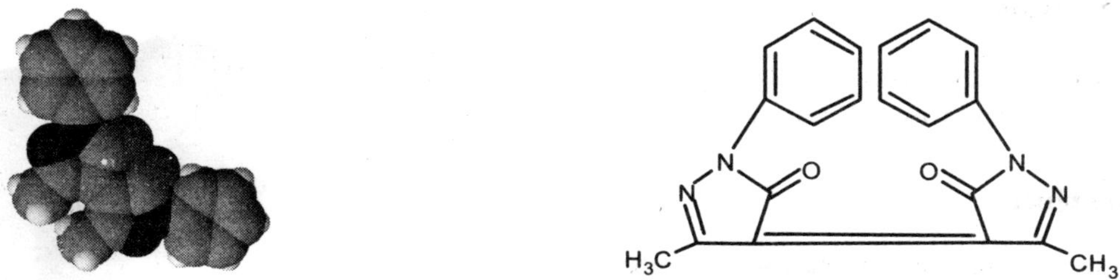

The imine structure is only found in the particular group od phenyl – methyl – pyrazolone derivative known as antipyrine.

Proof of the imine structure of antipyrine is found in the disruption of antipyrine by means of sodium and carbon dioxide, to give the anilide of β - methylamino – crotonic acid.

Three other groups of compounds must be considered an belonging to the phenolic type viz, the phenol – ethers, esters and salts of phenyl – methyl – pyrazolone.

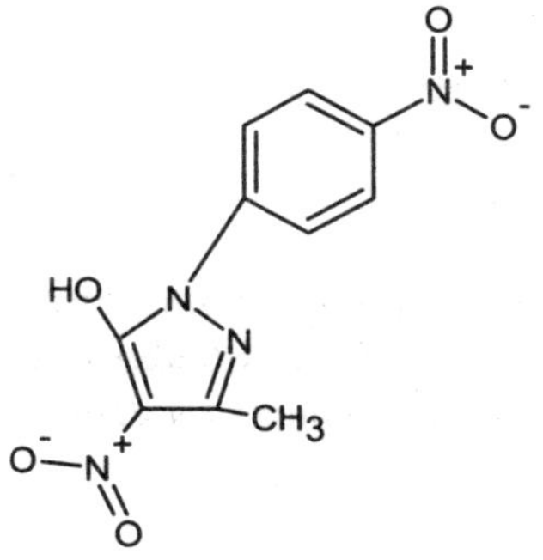

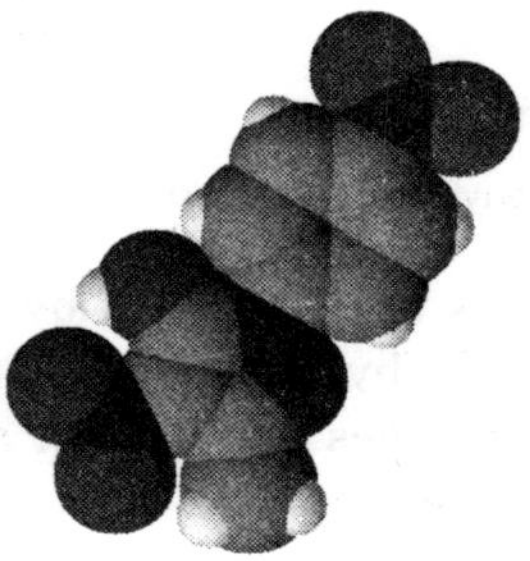

In conclusion, it may be mentioned that certain nitro – derivatives pyrazolone, known as picrolonic acid are also regarded by Knorr as nitro phenols owing to their similarity to picric acid. Picrolonic acid, yields very sparingly soluble salts which in their properties show a close resemblance to the picrates. They are usually even less soluble than the latter, and may be employed with advantage for the isolation and identification of bases.

The arguments which lends support to each of the three competing formula of phenyl – methyl pyrazolone lead to the conclusion that the acidic hydrogen atom and the double bonds occupy no fixed positions in this compound. It is the peculiar mobility of this hydrogen atom which enables tautomeric changes to the completed with such ease.

Analogous rearrangement of bonds but without any movement of hydrogen, are also shown by antipyrine in certain addition reactions.

ANTIPYRINE

Introduction

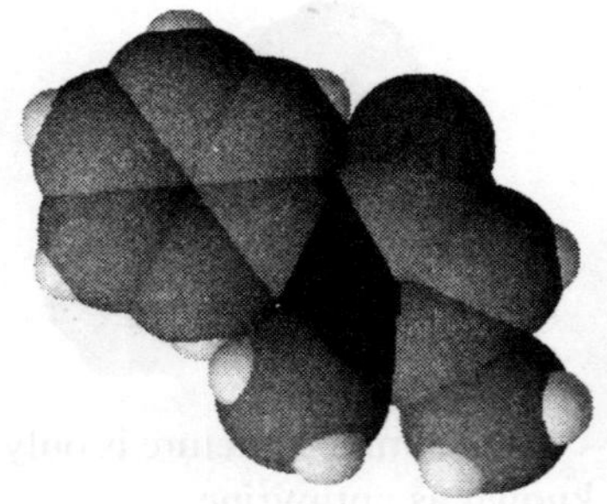

It is also known as phenazone, it is the most important member of the pyrazole group, and it is use extensively in medicine as a febrifuge*.

Preparation

It is prepared industrially by heating 1 – phenyl – 3 – methyl - 5 - pyrazolone with methyl iodide and methyl alcohol under 1 atmospheric pressure and 100°C temperature in an autoclave, thus after this the hydroiodate of the antipyrine is produces and which is reacted with sodium hydroxide to liberate antipyrine.

HO, N, N, CH_3 — methyl iodide /methyl alcohol, 100°C/ 1atm → O, N, N, CH_3, I, H — NaOH → O, N, N, H_3C, CH_3

3-methyl-1-phenyl-1H-pyrazol-5-ol — antipyrine hydroiodide — antipyrine

Antipyrine is also prepared by reducing diortho-dinitrodiphenyl with sodium amalgam and methyl alcohol, or by heating diphenylene-ortho-dihydrazine with hydrochloric acid to 150 C. It crystallizes in needles which melt at 156 C. Potassium permanganate oxidizes it to pyridazine tetracarboxylic acid.

*anti fever.

Physical Properties

Molecular Formula	$= C_{11}H_{12}N_2O$
Formula Weight	= 188.22578
Melting Point	=156 °C
Composition	= C(70.19%) H(6.43%) N(14.88%) O(8.50%)
Molar Refractivity	= 54.55 cm^3
Molar Volume	= 162.7 cm^3
Parachor	= 416.1 cm^3
Index of Refraction	= 1.585
Surface Tension	= 42.7 dyne/cm
Density	= 1.156 g/cm^3
Polarizability	= 21.62 $10^{-24}cm^3$
Monoisotopic Mass	= 188.094963 Da
Nominal Mass	= 188 Da
Average Mass	= 188.2258 Da
Log P	= 0.27

Chemical Properties

Antipyrine is both water and alcohol soluble. The aqueous solution is red in ferric chloride and green in nitrous acid.

HNO_2

H_3C CH_3 antipyrine → H_3C CH_3 4 - nitroso antipyrine

Antipyrine is a strong monoacid base and readily forms salts, most of which are easily soluble in water.

At ordinary temperatures alkyl iodides unite with antipyrine in the same manner as with the inner salts of phenol ammonium bases, i.e. *phenol betaines* in that iodine attaches itself to the 2 nitrogen atom of antipyrine, while the alkyl group unites with the oxygen atom. This behaviour originally led to the suggestion that antipyrine should be formulated as phenol – betaines.

From antipyrine and methyl iodide for example, there is formed an antipyrine "pseudo' methyl iodide which is identical with the methiodide of 1 phenyl 3 methyl -5 – methoxy pyrazole.

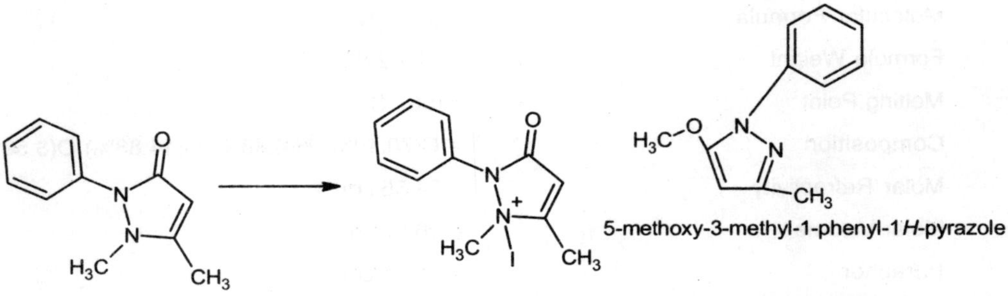

5-methoxy-3-methyl-1-phenyl-1*H*-pyrazole

1-iodo-1,5-dimethyl-3-oxo-2-phenyl-2,3-dihydro-1*H*-pyrazol-1-ium

1,5-dimethyl-2-phenyl-1,2-dihydro-3*H*-pyrazol-3-one

Phenol betaines unites with methyl iodide in a similar manner, in the presence of caustic potash. The resulting quaternary iodides are identical with those obtained from the dimethyl amino – anesoles.

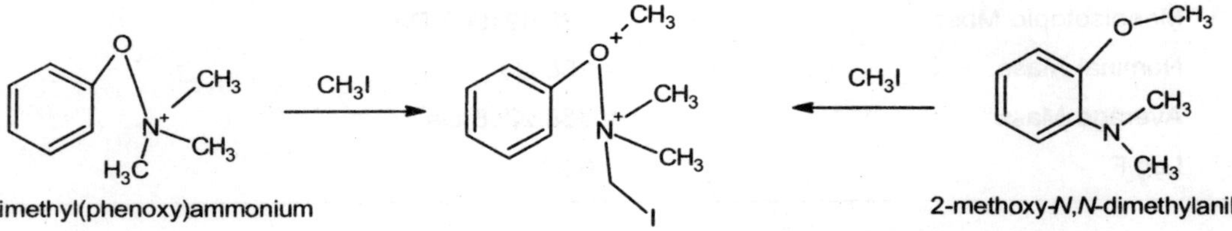

CH_3I CH_3I

trimethyl(phenoxy)ammonium

2-methoxy-*N*,*N*-dimethylaniline

[(iodomethyl)(dimethyl)ammonio](methyl)phenyloxonium

A detail comparison of antipyrine with o – trimethyl – ammonium phenoxides, however, revealed the fact that apart from the addition of alkyl iodide, these compounds were quite different in behaviour the pseudo – alkaloids of both compounds had also alkaloids of both compounds had also different properties, and hence the "phenol betaines" formula for antipyrine was rejected.

On being fused, the pseudo alkaloids of antipyrine do not break up as might be expected, into alkyl iodide and phenolic ester, but into alkyl iodide and antipyrine

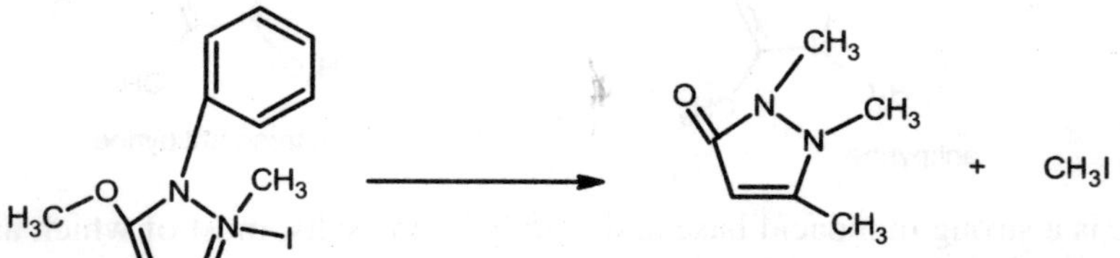

\+ CH_3I

1,2,5-trimethyl-1,2-dihydro-3H-pyrazol-3-one

2-iodo-5-methoxy-2,3-dimethyl-1-phenyl-1H-2λ-pyrazole

Antipyrine is also regenerated from the pseudo – methiodide by the action of alkalies, slowly in the cold and more rapidly on heating.

In the above reactions it will be seen that antipyrine behaves as an unsaturated compound of type I under the influence of alkalies at higher temperature, however antipyrine and similarly constituted compounds react as unsaturated substance according to formula II and III

Finally, on interaction with bromine, antipyrine behaves in the sense of formula IV.

(I) (II)

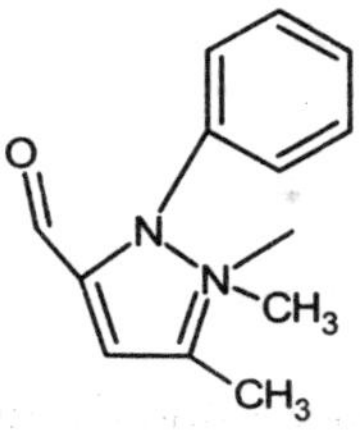

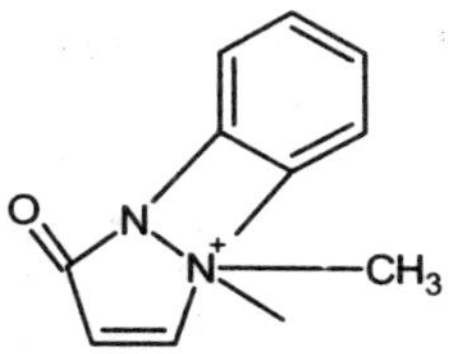

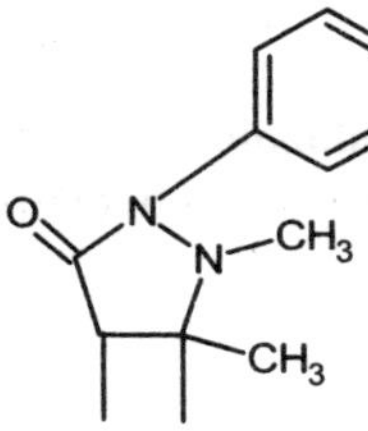

(III)

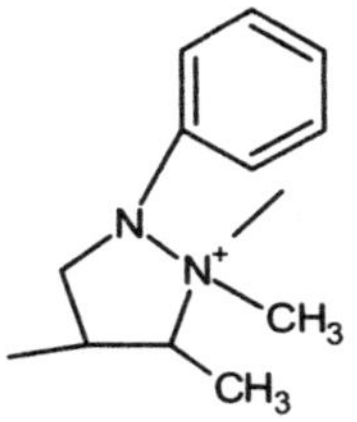

This remarkable variation in addition reaction is explained due to the intermolecular movements of the hydrogen atom in antipyrine, accomplished by change of linking.

It is suggested that the 2 – nitrogen atoms alternates between the tri and pentavalent states, thus permitting displacements of valency bonds similar to those assumed in the case of tautomeric compounds.

Among the great number of antipyrine molecules present in solution under the fused state there will be some in which the valency conditions corresponds to the above four types.

In the addition of alkyl halides in the cold, and in the production of the antipyrine salts, from I will react in preference, since the negative radical will naturally attach itself to the basic and the positive radical to the acidic point of the antipyrine molecule. Only at higher temperature, at which the alkaloids of the phenol – either type are no longer capable of existence does antipyrine react with methyl iodide according to formulæ II and III. Bromine, on the other hand unites in position 3 and 4 (from IV), in accordance with its tendency to add on to a double carbon linkage.

Certain other derivatives of this type are of importance, e.g. solipyrine, or antipyrine salicylate, tolipyrine or p – toly – dimethyl – pyrozolone and pyramidone or 4 – dimethyl amino – antipyrine. These and other derivatives are employed medicinally, particularly as substitutes for antipyrine.

Indazoles or Benzo – Pyrazoles

The rings systems of the indazoles consists of a condensed benzene pyrazoles nucleus and these compounds thus stand in the same relationship, to the pyrazoles as the indoles to the pyrazoles. Although the imino – hydrogen atom in indazoles is not definitely located on either nitrogen atom, the parent compound gives rise to two series of N – derivatives e.g. the 1 and 2 – N methyl indazoles. The position of substituents is indicated by numbers as in the following formula.

Indazole (A)   Idazole (B)

The structure of indazole has been intensively studied by Auwers and Co workers and is still a matter of controversy. In all probability it is a resonance hybrid of forms A and B with predominating A.

INDAZOLE

Introduction

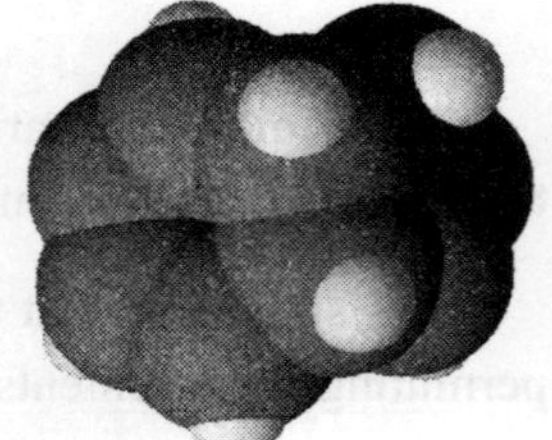

Indazole, also called benzpyrazole or Isoindazone is a heterocyclic aromatic organic compound.

Preparation

It is prepared from benzo – toluidide by dissolving it in acetic acid and acetic anhydride and passing in nitrous fumes; the resulting nitroso– compounds, when heated in dry benzene loses water and benzoic acid to form indazole, which is extracted with hydrochloric acid and then precipitated by addition of alkali. Indazole is soluble in water

N-phenylbenzamide → *N*-(2-methylphenyl)-*N*-nitrosobenzamide → Indazole + benzoic acid + H_2O

It is also obtained by removing the elements of water from o – toluene diazo – hydroxide in neutral solution

1-hydroxy-2-phenyldiazene → 1*H*-indazole + H_2O

A large number of indazoles substituted in the benzene nucleus have been prepared by this method by starting from substituted o – toluene diazo – hydroxides. The diazo compounds prepared from nitrated and brominated o – toluidines have a strong tendency to form rings of this type.

Physcial Properties

Molecular Formula	= $C_7H_6N_2$
Formula Weight	= 118.13594
Composition	= C(71.17%) H(5.12%) N(23.71%)
Melting point	= 147-149 °C
Boiling point	= 270 °C
Molar Refractivity	= 36.61 ± 0.3 cm^3
Molar Volume	= 95.0 ± 3.0 cm^3
Parachor	= 264.8 ± 4.0 cm^3
Index of Refraction	= 1.696 ± 0.02
Surface Tension	= 60.1 ± 3.0 dyne/cm
Density	= 1.242 ± 0.06 g/cm^3
Polarizability	= 14.51 ± 0.5 $10^{-24} cm^3$
Monoisotopic Mass	= 118.053098 Da
Nominal Mass	= 118 Da
Average Mass	= 118.1359 Da
Log P	= 1.82

IMINAZOLES OR GLYOXALINE

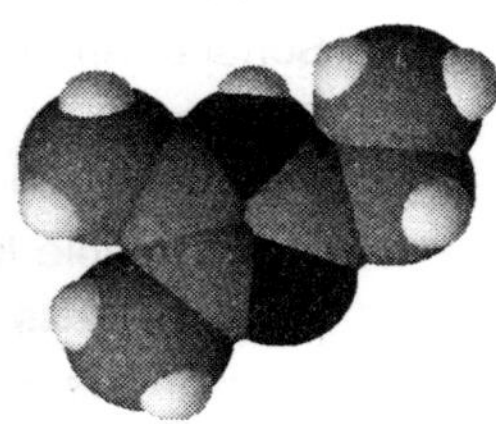

The ring system of the iminazoles, like that of the pyrazoles, consists of three carbon and two nitrogen atoms. In this case, however, the latter are not adjacent but are separated by a carbon atom. Hence iminazoles may be regarded as cyclic amindines, derived from the complex

$$HN = CH — NH_2$$

Iminazole, the parent compounds of the series, is formed by the reaction of ammonia on glyoxal is first broken up to give formic acid and formaldehyde, and that the latter than condenses with the ammonia and glyoxal:

NH_3 + oxalaldehyde + NH_3 + formaldehyde → Glyoxaline (positions: 1 (H)N, 2, N 3, (a)4, (b)5)

The figures or letters attached to the formula indicate the manner in which the substituents are represented.

The above is a somewhat trouble some method of preparing glyoxaline and it is more convenient to allow formaldehyde and excess of ammonia to interact with di nitro – tartaric acid, when as good yield of ammonium glyoxaline – dicarboxylate dicarboxylic acid is precipitated. On heating this to about 300°C it decomposes smoothly into carbon dioxide and glyoxaline.

Substituted glyoxalines are prepared by method analogous to that employed for glyoxalines, by the action of ammonia and aldehydes on glyoxal or other 1:2 diketo compounds.

$$RCOCOR + 2NH_3 + R^1CHO \longrightarrow \text{(substituted glyoxaline)} + 3H_2O$$

Glyoxaline forms prisms. It is weak base and when warm possesses a faint fishy smell.

Molecular Formula	$= C_7H_{12}N_2$
Formula Weight	= 124.18358
Composition	= C(67.70%) H(9.74%) N(22.56%)
Molar Refractivity	= 37.97 cm^3
Molar Volume	= 126.3 cm^3
Parachor	= 312.9 cm^3
Index of Refraction	= 1.513
Surface Tension	= 37.6 dyne/cm
Density	= 0.982 g/cm^3
Polarizability	= 15.05 $10^{-24}cm^3$
Monoisotopic Mass	= 124.100048 Da
Nominal Mass	= 124 Da
Average Mass	= 124.1836 Da

Chemical Properties

Glyoxalines are stronger bases then the isomeric pyrazoles, as many been seen from their constants (glyoxaline 1.2×10^{-7}) and pyrazole 3.0×10^{12}.

The imino – hydrogen atom of glyoxaline can be replaced by metals and alkyl radicals. Ammonical silver solutions yield with glyoxaline flocculent precipitate of the silver salt, which is only slightly soluble in excess of ammonia. The parent glyoxaline bases are quite stable towards alkalies, but this property is lost in the methiodides and similar derivatives, which on heating with alkalies decompose into mixture of two primary bases.

3-bromo-1-methyl-3-pentyl-1H-3λ[5]-imidazole → H_2N-CH_3 (methanamine) + $H_3C-CH_2-CH_2-CH_2-CH_2-NH_2$ (pentan-1-amine)

Glyoxaline and its simple substituents products in which the imino group is still intact are very readily disrupted, even at 0°C, by mixture of bezoyl chloride and caustic soda with the production of a carboxylic acid and a dibenzoylated base. Benzene diazo – chloride also reacts with glyoxalines containing a free imino hydrogen atom, to yield coloured diazo amino compounds.

Occurrence of Glyoxalanes in Some Natural Products

It has been shown by Pinner that the alkaloid pilocarpine, present in jaborandi leaves (of Pilocarpus perinantifolius) is a derivative, such a caffeine, Theoborane and theophyline have already been described under the purine group. The purines in fact contain a nucleus formed by the fusion of an iminazole ring with a pyremidine ring..

Iminiazole derivatives are also met with among the disruption products of the protein. The discovery of a remarkable transformation from the sugars to the iminazole group lend additions interest to this series from the physiological as well as the chemical stand point.

HISTAMINE

Histamine is a biogenic amine involved in local immune responses as well as regulating physiological function in the gut and acting as a neurotransmitter. New evidence also indicates that histamine plays a role in chemotaxis of white blood cells.

Synthesis and Metabolism

Histamine is derived from the decarboxylation of the amino acid histidine, a reaction catalyzed by the enzyme L-histidine decarboxylase. It is a hydrophilic vasoactive* amine.

Conversion of histidine to histamine by histidine decarboxylase

Once formed, histamine is either stored or rapidly inactivated. Histamine released into the synapses is broken down by acetaldehyde dehydrogenase. It is the deficiency of this enzyme that triggers an allergic reaction as histamines pool in the synapses. Histamine is broken down by histamine-N-methyltransferase and diamine oxidase. Some forms of foodborne disease, so-called "food poisonings," are due to conversion of histidine into histamine in spoiled food, such as fish.

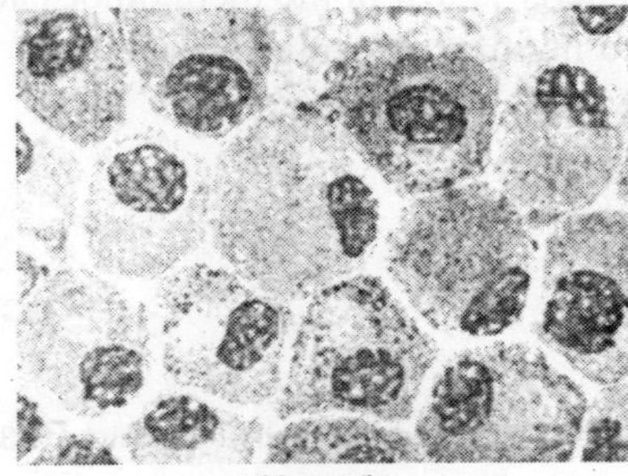

Mast cells

Storage and Release

Most histamine in body tissue is found in granules in mast cells (see figure) or basophils. Mast cells are especially numerous at sites of potential injury - the nose, mouth, and feet; internal body surfaces; and blood vessels. Non-mast cell histamine is found in several tissues, including the brain, where it functions as a neurotransmitter. Another important site of histamine storage and release is the enterochromaffin-like (ECL) cell of the stomach.

The most important pathophysiologic mechanism of mast cell and basophil histamine release is immunologic. These cells, if sensitized by IgE antibodies attached to their membranes, degranulate when exposed to the appropriate antigen. Certain amines, including such drugs as morphine and tubocurarine, can displace histamine in granules and cause its release.

Mechanism of Action

Histamine exerts its actions by combining with specific cellular receptors located on cells. The four histamine receptors that have been discovered are designated H_1 through H_4.

Type	Location	Function
H_1 histamine receptor	Found on smooth muscle, endothelium, and central nervous system tissue	Causes vasodilation, bronchoconstriction, smooth muscle activation, separation of endothelial cells (responsible for hives), and pain and itching due to insect stings; the primary receptors involved in allergic rhinitis symptoms and motion sickness.

*acts on the contraction and relaxation of the blood vessels.

Type	Location	Function
H_2 histamine receptor	Located on parietal cells	Primarily stimulate gastric acid secretion
H_3 histamine receptor	-	Decreased neurotransmitter release: histamine, acetylcholine, norepinephrine, serotonin
H_4 histamine receptor	Found primarily in the thymus, small intestine, spleen, and colon. It is also found on basophils and in the bone marrow.	Unknown physiological role.

Use

Histamine Helps Wound Healing. When you hear the word histamine do you begin tearing and sneezing? Millions do but what about your *skin*? In an earlier study intended to explore the role of histamine in animals one finding related to the delayed healing of skin at wound sites. In this follow-up study (Japanese Society for Investigative Dermatology Annual Meeting, Ehime, Japan, September 7-8, 2001) genetically modified mice unable to produce histamine were found to show substantially protracted healing of wounds. Application of a histamine-containing solution to the wounds not only normalised wound healing but also accelerated it beyond that of the normal unmodified mice. Additional investigations led to the conclusion that histamine promoted new blood vessel growth (angiogenesis), an absolutely critical component of wound healing. This may explain why the dipeptide L-carnosine, which contains L-histidine (the direct precursor of *histamine*), is an effective wound healing promoter and the subject of new patents.

BENZIMINAZOLES OR BENZO GLYOXALINES

These compounds condensed glyoxaline benzene structure, and bear to the glyoxalines the same relationship as the indoles to the pyrroles. They are cyclic o – amindines of the benzene series, and are formed by the condensation of o – phenylene diamine and its substitution products with carboxylic acid or their anhydrides.

benzene-1,2-diamine + formic acid → 1*H*-benzimidazole + $2H_2O$

Physcial Properties

Molecular Formula	= $C_7H_6N_2$
Formula Weight	= 118.13594
Composition	= C(71.17%) H(5.12%) N(23.71%)
Molar Refractivity	= 36.61 cm^3
Molar Volume	= 95.0 cm^3
Parachor	= 264.8 cm^3
Index of Refraction	= 1.696
Surface Tension	= 60.1 dyne/cm
Density	= 1.242 g/cm^3
Polarizability	= 14.51 ± 0.5 $10^{-24}cm^3$
Monoisotopic Mass	= 118.053098 Da
Nominal Mass	= 118 Da
Average Mass	= 118.1359 Da
Log P	= 1.38

Chemical Properties

The basic character of the benziminazoles is not quite so marked as that of the glyoxalines. They are also weakly acidic and generally soluble in aqueous alkalies with the formation of N – metallic compounds.

ISOXYAZOLES AND OXALOES AND THIAZOLES

In those rings composed of three carbons linked with one nitrogen atoms of carbon. In the former case the compounds are termed as isoxazoles and later oxazoles.

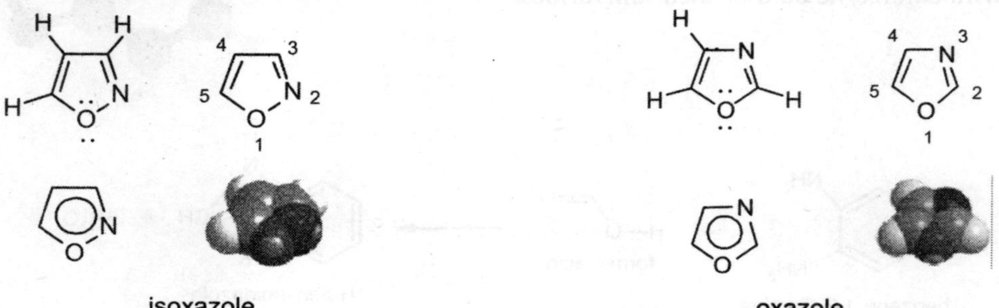

isoxazole

oxazole

Isoxazole correspond to pyrazoles and just as the latter are obtained from hydrazones, the former result by loss of water from the monoximes of β - diketones and β - ketoaldehydes.

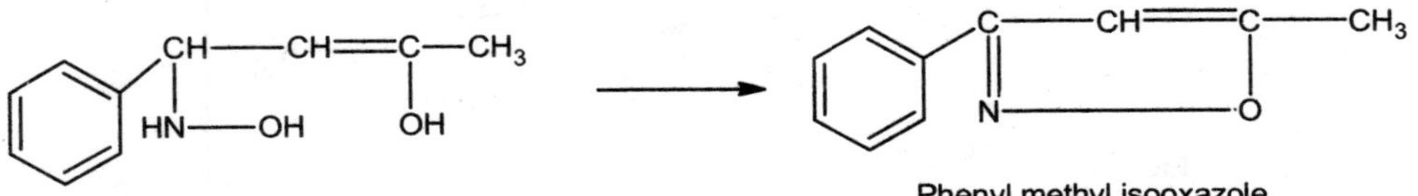

Monoxime of aci - benzoyl acetone

Phenyl methyl isooxazole

Preparation

The **Robinson-Gabriel synthesis** is a chemical reaction that forms oxazoles by dehydration of 2-acylamino-ketones.

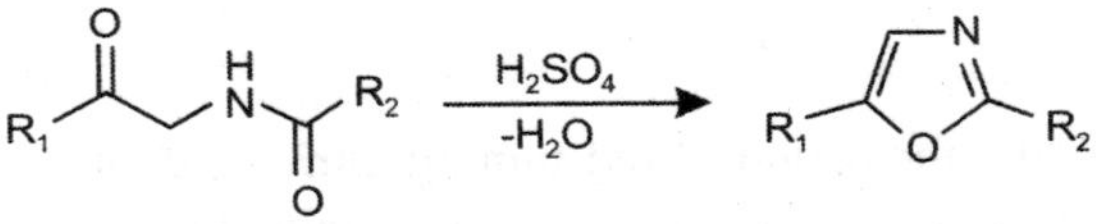

H_2SO_4
$-H_2O$

Oxazolines can also be obtained from cycloisomerization of certain propargyl amides. In one study oxazoles are synthesized in a one-pot synthesis starting with condensation of propargyl amine and benzoyl chloride to the amide followed by Sonogashira coupling of the terminal alkyne end with another equivalent of benzoylchloride and concluded with *p*-toluenesulfonic acid catalyzed cycloisomerization:

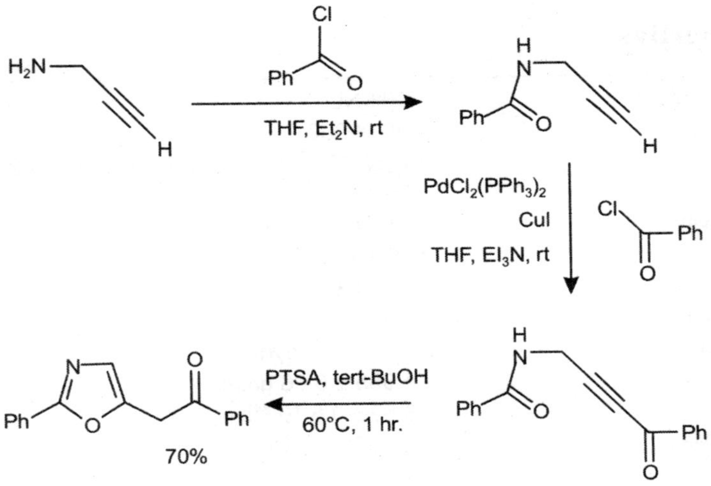

THF, Et_3N, rt

$PdCl_2(PPh_3)_2$
CuI
THF, El_3N, rt

PTSA, tert-BuOH
60°C, 1 hr.

70%

In one reported oxazole synthesis the reactants are a benzoyl chloride and an isonitrile

They are also produced by the action of red fuming nitric acid on diketones and esters of ketonic acid. The parent compound of the group, isooxazole, is obtained by the interaction of hydroxyl amine and propagylic aldehyde. Assuming that an oxime is first produced, this reaction may be considered as an inter molecular addition of the oximino – group, N OH to the triple carbon linking.

acrylaldehyde

N-hydroxyallen-1-amine

3H-isoxazol-1-ium

Physical Properties

Molecular Formula	= C_3H_3NO
Formula Weight	= 69.06202
Composition	= C(52.17%) H(4.38%) N(20.28%) O(23.17%)
Boiling point	= 95 °C
Molar Refractivity	= 16.64 ± 0.3 cm^3
Molar Volume	= 65.4 ± 3.0 cm^3
Parachor	= 155.4 ± 4.0 cm^3
Index of Refraction	= 1.422 ± 0.02
Surface Tension	= 31.7 ± 3.0 dyne/cm
Density	= 1.055 ± 0.06 g/cm^3
Polarizability	= 6.59 ± 0.5 $10^{-24} cm^3$
Monoisotopic Mass	= 69.021464 Da
Nominal Mass	= 69 Da
Average Mass	= 69.062 Da
Log P	= 0.08

Chemical Properties

Oxazoles and isoxazoles like pyrazoles are weak bases. On being mixed with and alcoholic solution of sodium ethoxide, they decompose to give the sodium salt of cyano – acetaldehyde or cyano – vinyl alcohol.

1,3-oxazole + $NaOC_2H_5$ (sodium ethaoxide) ⟶ sodium (*E*)-2-cyanoethenolate + H_3C–CH_2–OH

In an analogous manner α - alkyl – isoxazoles , in which the γ - position is unsubstituted are attached comparatively rapidly by alkalies and instantaneously by sodium ethoxide forming salts of the isomeric cyano – ketones.

5-methylisoxazole ⟶ 3-oxobutanenitrile

On the other hand, α - γ dialkyl isoxazoles are extremely stable towards alkalies.

Uses

Oxazoles are used in Alkaloid Synthesis.

THIAZOLE

Introduction

Thiazole, or 1,3-thiazole, is a clear to pale yellow flammable liquid and pyridine-like odor with the molecular formula C_3H_3NS. It is a 5-membered ring, in which two of the vertices of the ring are nitrogen and sulphur, and the other three are carbons.

Preparation

Thiazoles are formed by the interaction of thiamides and α - chloroketones or α - chloro – aldehydes.

1-chloroacetone + ethanimidothioic acid ⟶ 4-methyl-1,3-thiazole + hydrogen chloride (HCl) + water (H_2O)

If thiourea is used in this reaction μ - amino – Thiazoles formed which on treatment with nitrous acid and alcohol exchange the amino group from hydrogen, with the production of Thiazoles.

Thiazoles itself is prepared by the above method from μ - amino Thiazole and it forms a mobile, volatile liquid boils at 117°C and it smell like pyridine. It is less basic than the latter. A large number of derivatives of thiazole are known, which cannot de described here

Benzo thiazole resembles the quinoline bases, and corresponds in their composition to the benzoxazoles and benziminazoles. They are produced by the action of acids on – amino thio phenols.

2-aminobenzenethiol + acetic acid ⟶ 2-methyl-1,3-benzothiazole + $2H_2O$

Certain derivatives of this group are of value as substance cotton dyes.

Thus when ρ – toluidine and sulphur are heated together for a considerable at 200°C "primuline base" is obtained, containing the following thiazole derivative.

3-(6-methyl-2,6'-bi-1,3-benzothiazol-2'-yl)aniline

Physcial Properties

Molecular Formula	= C_3H_3NS
Formula Weight	= 85.12762
Boiling point	= 116-118 °C
Composition	= C(42.33%) H(3.55%) N(16.45%) S(37.67%)
Molar Refractivity	= 22.72 cm^3
Molar Volume	= 72.1 cm^3
Parachor	= 184.5 cm^3
Index of Refraction	= 1.542
Surface Tension	= 42.8 dyne/cm
Density	= 1.180 g/cm^3
Polarizability	= $9.01 \pm 0.5\ 10^{-24} cm^3$
Monoisotopic Mass	= 84.998619 Da
Nominal Mass	= 85 Da
Average Mass	= 85.1276 Da
Log P	= 0.44

Chemical Properties

Thiazoles are characterized by larger pi-electron delocalization than the corresponding oxazoles and have therefore greater aromaticity. This is evidenced by the position of the ring protons in proton NMR (between 7.27 and 8.77 ppm), clearly indicating a strong diamagnetic ring current.

The calculated pi-electron density marks C5 as the primary electrophilic site, and C_2 as the nucleophilic site.

The reactivity of a thiazole can be summarized as follows:

- Deprotonation at C2: the negative charge on this position is stabilized as an ylide; Grignard reagents and organolithium compounds react at this site, replacing the proton

2-(trimethylsiliyl)thiazole [5] (with a trimethylsilyl group in the 2-position) is a stable substitute and reacts with a range of electrophiles such as aldehydes, acyl halides, and ketenes

- Alkylation at nitrogen forms a thiazolium salt
- Electrophilic aromatic substitution at C5 requires activating groups such as a methyl group in this bromination:

Br_2

- Nucleophilic aromatic substitution often requires an electrofuge at C2, such as chlorine with

$NaOCH_3$

- Organic oxidation at nitrogen gives the thiazole N-oxide; many oxidizing agents exist, such as mCPBA; a novel one is hypofluorous acid prepared from fluorine and water in acetonitrile; some of the oxidation takes place at sulphur, leading to a sulfoxide [6]:

$$F_2 + H_2O + MeCn \longrightarrow HOF.MeCN$$

HOF.MeCN, CH_2CL_2

95% 5%

- Thiazoles are formyl synthons; conversion of **R-thia** to the **R-CHO** aldehyde takes place with[5], respectively, methyl iodide (N-methylation), organic reduction with sodium borohydride, and hydrolysis with mercury chloride in water.
- Thiazoles can react in cycloadditions, but in general at high temperatures due to favourable aromatic stabilization of the reactant; Diels-Alder reactions with alkynes are followed by extrusion of sulphur, and the endproduct is a pyridine; in one study [4], a very mild reaction of a *2-(dimethylamino)thiazole* with *dimethyl acetylenedicarboxylate* (DMAD) to a pyridine was found to proceed through a zwitterionic intermediate in a formal [2+2]cycloaddition to a cyclobutene, then to a *1,3-thiazepine* in an 4-electron electrocyclic ring openening and then to a *7-thia-2-azanorcaradiene* in an 6-electron electrocyclic ring, closing before extruding the sulphur atom.

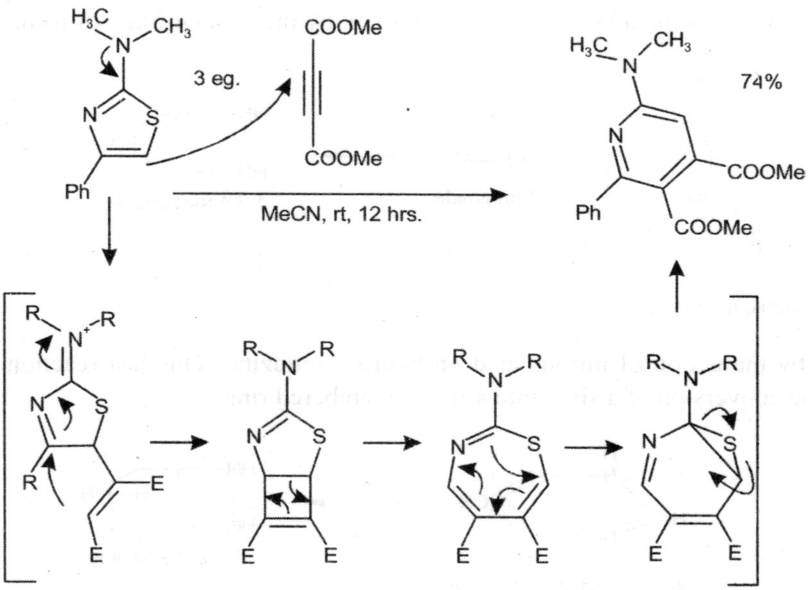

H$_3$C CH$_3$ N COOMe 3 eg. N S Ph COOMe MeCN, rt, 12 hrs. H$_3$C CH$_3$ N 74% N COOMe Ph COOMe

R N$^+$ R N S R E E R N R N S E E R N R N S E E R N R N S E E

Uses

Thiazole is used for manufacturing biocides, fungicides, pharmaceuticals, and dyes.

TRIAZOLES OR PYRODIAZOLES

If two of the CH groups in pyrrole are replaced by two N atoms four different ring systems may be derived as represented as following.

I II III IV

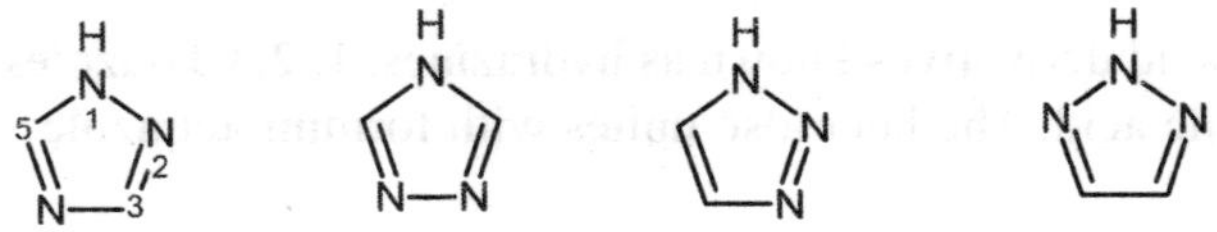

1,2,3 triazole 1,2,5 triazole

In this series we meet tautomeric phenomena recalling those described under pyrazole, where as all four compounds are known in the form of their N – alkyl and N – aryl derivatives, the parent substance I appears to be identical with II and similarly III with IV. For this reason it is convenient to make use of the following formulæ

Sym triazole v - triazole or oso triazole

in which the position of the mobile hydrogen atom is not specified.

Sym – triazole is obtained by various reactions e.g. by the condensation of formyl hydrazide with fromamide.

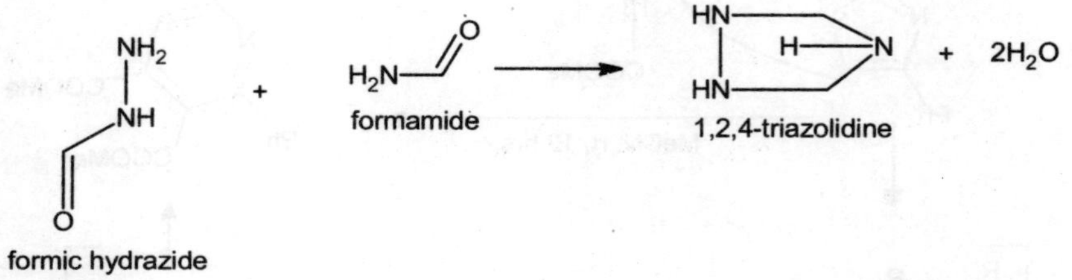

formic hydrazide + formamide → 1,2,4-triazolidine + $2H_2O$

And also by the action of nitrous acid on hydro – tetrazine. This last reaction is of interest as illuminating the conversion of a six – into a five – membered ring.

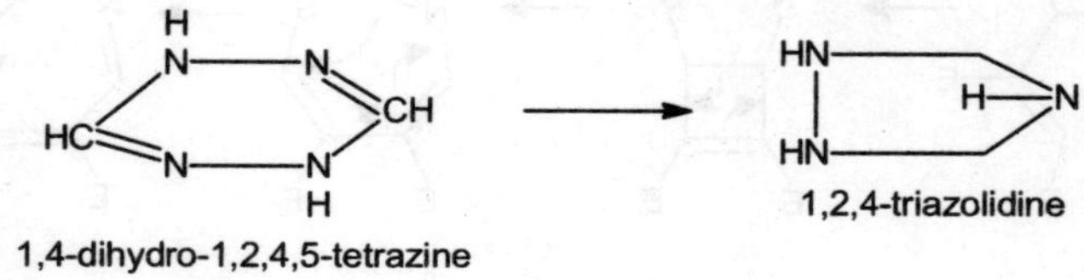

1,4-dihydro-1,2,4,5-tetrazine → 1,2,4-triazolidine

Both these methods are general application and symmetrical or true triazole can therefore be prepared,

1. From Dihydro – tetrazines

2. By the action of acid hydrides on amides.

Instead of starting from the hydrazide itself, the hydrazine hydrochloride proportions of the amide. In this case ammonia is first liberated with the formation of a hydride, which then act upon the second molecule of amide.

3. By warming the acid derivatives known as hydrazines. 1, 2,3 Triazoles are formed by the action of acetylene on hydrazoic acid. The later also unites with forming tetrazole.

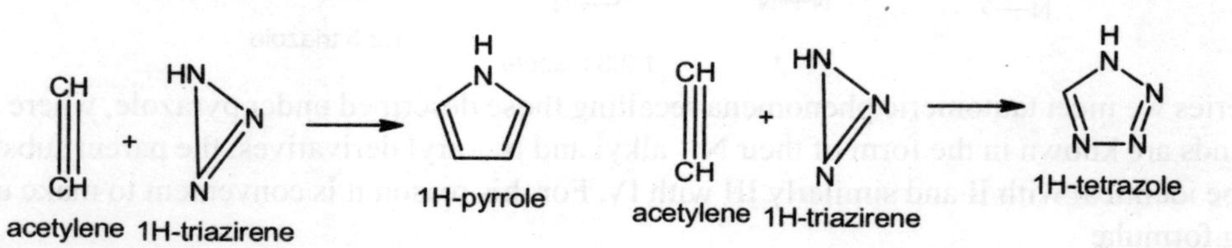

acetylene + 1H-triazirene → 1H-pyrrole; acetylene + 1H-triazirene → 1H-tetrazole

Phenylazide and sodium ethoxide in boiling alcoholic solution give I – phenyl 1: 2: 3 triazole as the main product of reaction. The use of diazo benzene imide and higher alkoxides lead to in general formation of 4 – alkyl – I phenyl 1:2:3 triazole.

Sulphur derivatives of triazole are produced by the action of phenyl – hydrazine persulphocyanic acid.

Persulphocyanic acid + $C_6H_5NHNH_2$, −S → 1-[(1-phenylhydrazino)methyl]thiourea, $-NH_3$ → Phenyl dithio triazolidone

Triazole sublimes in needles m.p 120°C to 121 and b.p 260°C, the physical properties are shown below.

Physical Properties

Molecular Formula	= $C_2H_3N_3$
Formula Weight	= 69.06532
Composition	= C(34.78%) H(4.38%) N(60.84%)
Molar Refractivity	= 16.86 ± 0.3 cm^3
Molar Volume	= 54.2 ± 3.0 cm^3
Parachor	= 155.1 ± 4.0 cm^3
Index of Refraction	= 1.534 ± 0.02
Surface Tension	= 67.1 ± 3.0 dyne/cm
Density	= 1.274 ± 0.06 g/cm^3
Polarizability	= 6.68 ± 0.5 $10^{-24}cm^3$
Monoisotopic Mass	= 69.032697 Da
Nominal Mass	= 69 Da
Average Mass	= 69.0653 Da
LogP	= 0.23

Chemical Properties

In general, the triazoles closely resemble the pyrazoles in behaviour, but are even more stable towards oxidizing agents. They are all weak base, although the introduction of two methyl or ethyl groups some what increases the basic strength. As in the case of the pyrazole, a number of interesting tautomeric phenomena have been discovered among the 1:2:3 triazoles.

Other ring compounds of similar type are the furazanes (I) resulting from the oximes of α- diketones by removal of water, and the oxydiazoles (II), obtained from symmetrical diacyl – hydrazines.

R OH N R N OH → R N O N R   HN R O HN R O → N R N O R

ENDIMINO – TRIAZOLES

Peculiar triazole bases containing a "nitrogen bridge, and known as endimino triazoles, have been prepared by Büsch. They are produced :

1. From acid chlorides and triaryl amino guanidines

propanoyl chloride + N,N',N"-trimethylcarbonohydrazonic diamide → 4-ethyl-3,5,6-trimethyl-2,3,5,6-tetraazabicyclo[2.1.1]hex-1-ene + H_2O + HCl

2. By condensing aldehydes with triaryl amino – guanidines, and oxidizing the resulting products.

1-ethyl-1-[(ethylamino)methyl]-3-methyltriaz-1-en-1-ium —$-H_2O$→ 4-ethyl-N,1, 5-trimethyl-4,5-dihydro-1H-1,2,4-triazol-3-amine —oxidation→ 1,4-dimethyl-2,3,6,7-tetraazabicyclo [3.1.1]hept-3-ene

Most of the endimino triazoles are yellow compounds, which possess strong basic character and crystallize well. Although very stable towards acids, they are readily decomposed into their original components by alkalies.

Endimino – triazoles are also of practical interest in so far their nitrates are much more sparingly soluble than any other nitrates yet examined, so that these bases may be employed as a reagent for the nitrate ion.

The nitrate of 1: 4 diphenyl endanilo – Dihydro – triazole is the least soluble of these compounds and may be used successfully for the qualitative and also the quantitative estimation of nitric acid, for these reasons this has been termed as nitron.

Nitron is prepared, from triphenyl – aminoguanidine and formic acid according to the following equation.

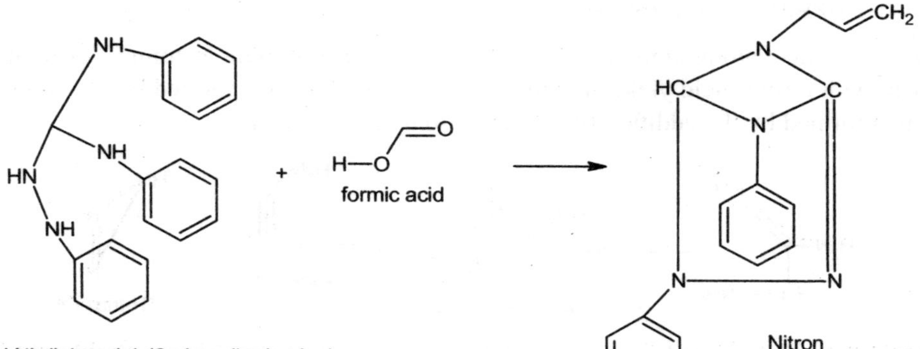

N,N'-diphenyl-1-(2-phenylhydrazino) methanediamine triphenyl amino guanidine + formic acid ⟶ Nitron

diphenyl endanilo dihydro triazole

As a reagent a 10% solution of nitron in 5% solution of acetic acid. The test sample is acidified with a drop of sulphuric acid and then five to six drops of nitron solution is added and a copious white precipitate immediately separates and by this means nitric acid can be detected even at the dilutions 1 to 60, 000. Nitron may also be employed for the detection of nitric acid in the presence of nitrites.

Schönberg has pointed out that the structure advanced by Büsch for the above compounds are very improbable owing to the distortion of the normal valency angles of the carbon joining the double bond to the four membered ring. He suggests the annexed formula of nitron.

TETRAZOLES

Tetrazoles contain a ring system built up from one carbon and four nitrogen atoms. Once again tautomeric phenomena are observed, similar to those described in the case of pyrazole and triazole, the one parent compound (I) giving rise to two series of derivatives (1a and 1b).

I Ia Ib

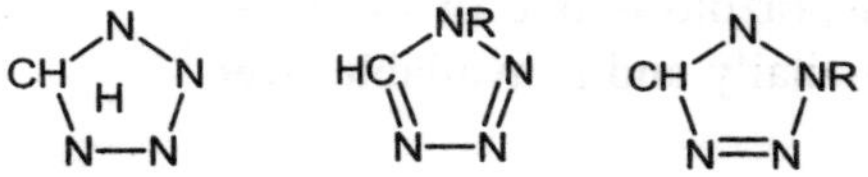

Owing to the mobility of the molecule double bonds, the hydrogen compound corresponding to the types Ia and Ib are in a state of dynamic equilibrium with one another, and isomerism can only be detected if intra molecular change is regarded by substituting the imino hydrogen atoms.

Tetrazoles are formed by various reasons e.g. with great readiness from nitrous acid and hydrazines.

(1Z)-ethanehydrazonamide + nitrous acid → 2*H*-tetrazole + $2H_2O$

An arrangement of atoms similar to that in the hydrazines is present in amino – guanidine. He latter on treatment with nitrous acid yields amino – tetrazole which is also formed by the amino – tetrazole which is also formed by the addition of cynamide to hydrazoic acid.

→ 1*H*-tetrazol-5-amine ← cyanamide + 1H-triazirene

Tetrazole may be obtained from amino-tetrazole by way of the diazo – compound in the same manner as benzene is obtained from aniline.

Tetrazole ring is very similar in nature to the benzene ring. Those tetrazole containing a free imino group are strong mono – basic acids. The silver and copper salts of tetrazole explode violently when heated.

Tetrazole forms colourless crystals of melting point 156°C and its aqueous solutions are acid in reaction. It possesses no basic properties and gives no nitro derivative.

Other Azoles

OMEPRAZOLE

Omeprazole is a proton pump inhibitor used in the treatment of dyspepsia, peptic ulcer disease (PUD), gastroesophageal reflux disease (GORD/GERD) and Zollinger-Ellison syndrome. It was first marketed by AstraZeneca as the magnesium salt **omeprazole magnesium** under the trade names **Losec** and **Prilosec**, and is now also available from generic manufacturers under various trade names. Omeprazole is one of the most widely prescribed drugs internationally and is available over the counter in some countries.

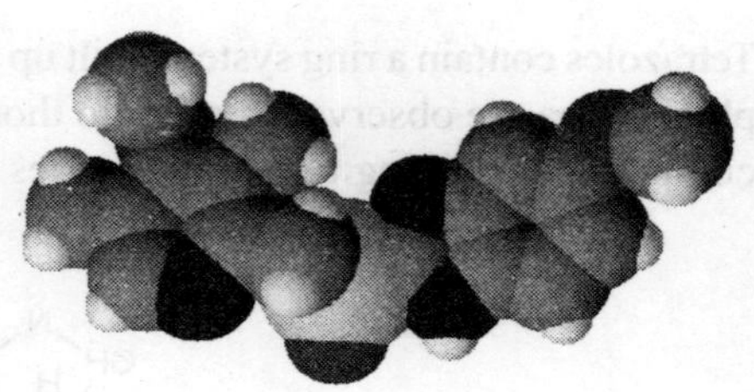

Facing the loss of patent protection and competition from generic manufacturers, AstraZeneca developed, launched, and heavily marketed esomeprazole (Nexium), a single enantiomer form of omeprazole. It has been proven that only the S-enantiomer is effective and the other is not. The type of cytochrome P450 enzyme that metabolizes the drug is 2C19, and the expression of this CYP450 enzyme varies between cultures, as elaborated below.

Omeprazole is the racemate (S and R enantiomers) undergoing a chiral shift in vivo from the racemate to the S enantiomer (active form). This chiral shift is accomplished by the CYP 2C19 Cytochrome which is not found equally in all human populations. Those who do not metabolize the drug effectively are called "poor metabolizers" and their distribution is as follows: - Caucasians 8% - 12% - Asian 20% - South Pacific Islands 70% Esomeprazole is the S enantiomer in the pure form, and its effects on the proton pump is therefore equal in all these populations, eliminating the "poor metabolizer effect".

In 1990, at the request of the United States Food and Drug Administration (FDA), the brand name *Losec* was changed to *Prilosec* to avoid confusion with the diuretic *Lasix* (furosemide). Unfortunately, the new name has led to confusion between omeprazole (*Prilosec*) and fluoxetine (*Prozac*), an antidepressant.

Use in *Helicobacter pylori* eradication

Omeprazole is combined with the antibiotics clarithromycin and amoxicillin (or metronidazole in penicillin-hypersensitive patients) in the one week eradication triple therapy for *Helicobacter pylori*. Infection by *H. pylori* is the causative factor in the majority of peptic and duodenal ulcers.

MICONAZOLE

Miconazole is an imidazole antifungal agent, developed by Janssen Pharmaceutica, and commonly applied topically (to the skin) or mucus membranes to cure fungal infections. It works by inhibiting the synthesis of ergosterol, a critical component of fungal cell membranes. It can also be used against certain species of Leishmania protozoa (which are a type of unicellular parasite), as these also contain ergosterol in their cell membranes. In addition to its antifungal and antiparasitic actions, it also has some limited antibacterial properties.

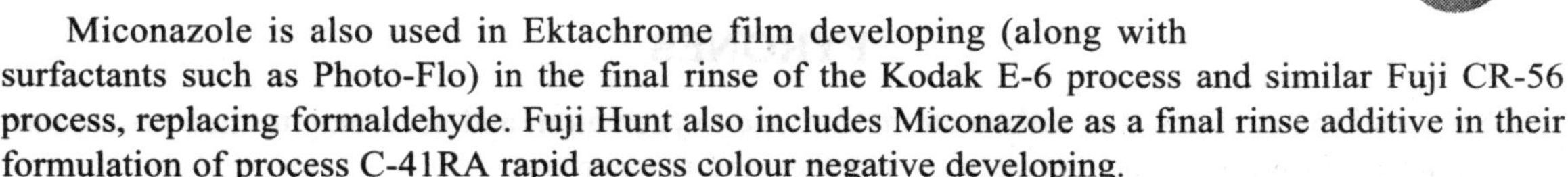

Miconazole is also used in Ektachrome film developing (along with surfactants such as Photo-Flo) in the final rinse of the Kodak E-6 process and similar Fuji CR-56 process, replacing formaldehyde. Fuji Hunt also includes Miconazole as a final rinse additive in their formulation of process C-41RA rapid access colour negative developing.

KETOCONAZOLE

History

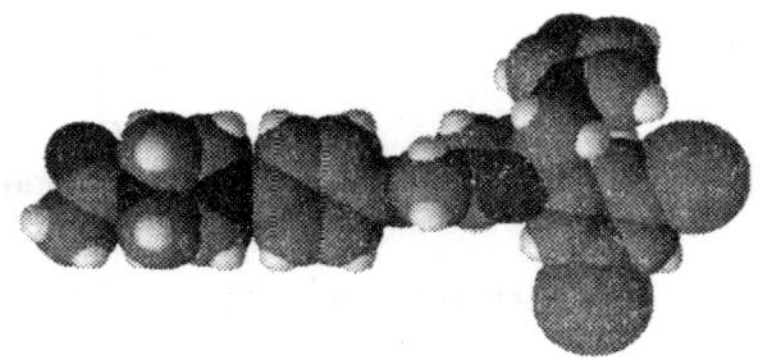

Ketoconazole was discovered in 1976 and released in the early 1980s, and was one of the first available oral treatment for fungal infections (griseofulvin was available before ketoconazole).

Ketoconazole is a synthetic antifungal drug used to prevent and treat skin and fungal infections, especially in immunocompromised patients such as those with AIDS. Due to its side-effect profile, it has been superseded by newer antifungals, such as fluconazole and itraconazole.[1] Ketoconazole is sold commercially as an anti-dandruff shampoo, branded **Nizoral®**, by Janssen Pharmaceutica.

Ketoconazole is very lipophilic, which leads to accumulation in fatty tissues. The less toxic and more effective triazole compounds fluconazole and itraconazole have largely replaced ketoconazole for internal use. Ketoconazole is best absorbed at highly acidic levels, so antacids or other causes of decreased stomach acid levels will lower the drug's absorption when taken orally.

Uses

Ketoconazole is usually prescribed for infections such as athlete's foot, ringworm, candidiasis (yeast infection or thrush), and jock itch. The over-the-counter shampoo version can also be used as a body wash for the treatment of tinea versicolor. Ketoconazole is used to treat eumycetoma, the fungal form of mycetoma. The side-effects of ketoconazole are sometimes used to treat non-fungal problems. The decrease in testosterone caused by the drug makes it useful for treating prostate cancer and for preventing post-operative erections[following penile surgery. Another use is the suppression of glucocorticoid synthesis, where it is used in the treatment of Cushing's disease. These side effects have also been studied for use in reducing depressive symptoms and drug addiction; however, it has not succeeded in either of these roles. Ketoconazole can be prescribed as a 200-mg pill, a 2% cream, a 2% gel, a 2% foam, or 2% shampoo for the treatment of dandruff or seborrhoeic dermatitis, or as a 1% over-the-counter shampoo (*Nizoral*).Ketokonazole is also available as a topical mousse, using patented Versafoam technology, marketed under the brand name Ketomousse. In clinical studies, the Versa foam proved to be a superior mechanism of delivery to the shampoo. Currently it is only available in Europe. The anti-dandruff shampoo is designed for people who have a more serious case of dandruff where symptoms include, but are not limited to constant non-stop flaking, and severe itchiness. The 2% shampoo is also being used as an additional hair loss treatment, and it has shown to have positive effects on androgenetic alopecia when used twice a week, mainly due to its anti-androgenic effects.

PYRONES

The pyrone ring contains five carbon atoms and one oxygen atom and according to their arrangement a distinct inference is drawn between γ - pyrones and α - pyrones.

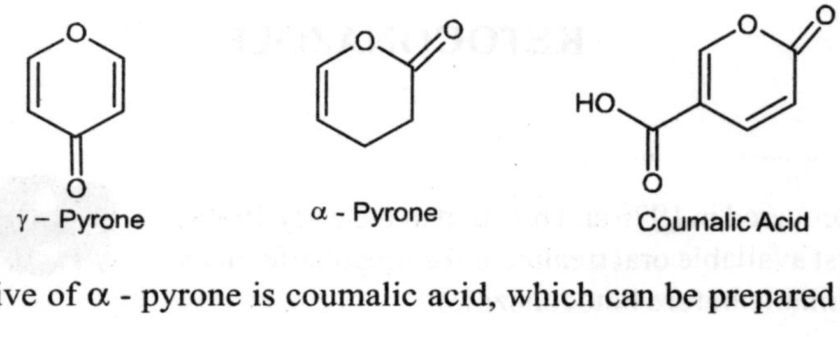

O O HO O O O O O

γ – Pyrone α - Pyrone Coumalic Acid

A simple derivative of α - pyrone is coumalic acid, which can be prepared from malic acid.

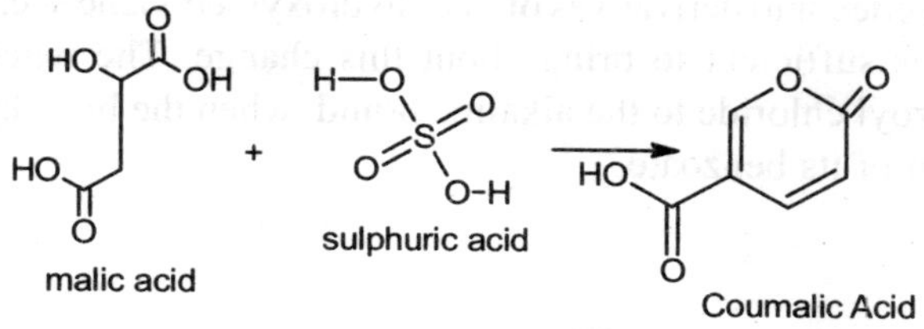

HO OH O HO O + H—O O=S=O O-H → HO O O O

malic acid sulphuric acid Coumalic Acid

In the following explanation of γ - pyrones which are generally called pyrones

γ – PYRONES

γ Pyrone is the parent compound of a series of substance found in nature e.g brazidin, the colouring matter of red – wood and in recent years the pyrone have also attached interest in connection with investigations on the basic properties of oxygen

A naturally occurring derivative of γ - pyrone is the meconic acid present in opium. On being heated, this acid parts with carbon dioxide to form comenic acid and finally pyromeconic acid (also called pyrocoumenic acid).

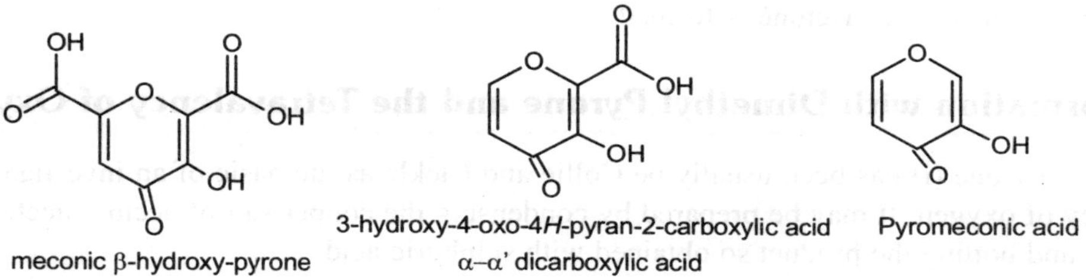

meconic β-hydroxy-pyrone 3-hydroxy-4-oxo-4*H*-pyran-2-carboxylic acid α–α' dicarboxylic acid Pyromeconic acid

Another compound of the type is chelidonic acid, found in the chelandine and white hellbore. On heating, this yields comanic acid and then pyrone.

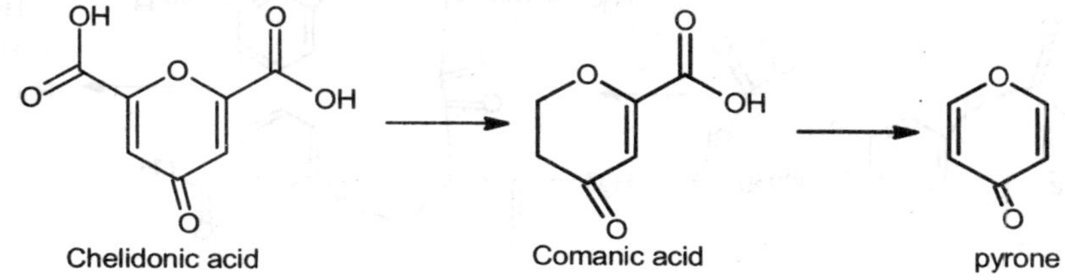

Chelidonic acid → Comanic acid → pyrone

Cheledonic acid and pyrone are easily disrupted to give open chain compounds. On being boiled with alkali, the former decomposes smoothly into 1 mol of acetone and 2 mole oxalic acid.

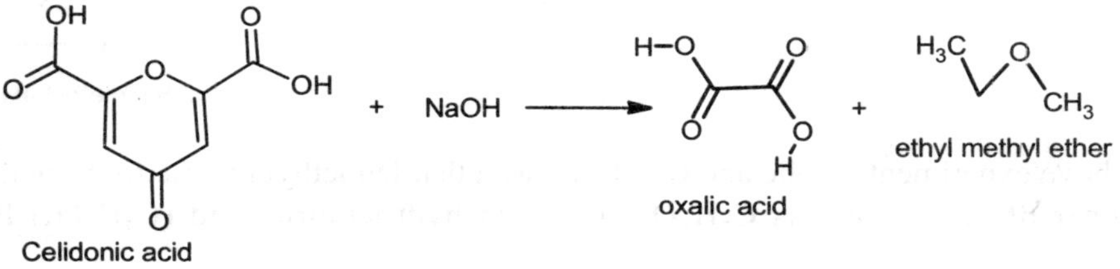

Celidonic acid + NaOH → oxalic acid + ethyl methyl ether

Pyrone is easily converted into derivatives of bis – hydroxymethylene acetone. Even a short treatment with alkali in the cold is sufficient to bring about this change. The reaction may be conveniently followed by adding benzoyl chloride to the alkaline liquid, when the bis – hydroxy methylene acetone separates out in the form of its benzoate.

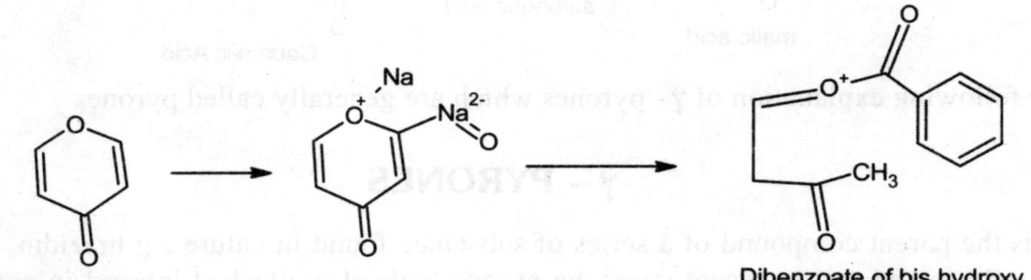

4*H*-pyran-4-one

Dibenzoate of bis hydroxy methylene acetone

Owing to its insolubility this dibenzoyl compound provides a useful means of testing for pyrone itself in dilute aqueous solution.

A similar opening of the pyrone ring is produced by the action of aniline acetate, when the dianilide of bis – hydro methylene acetone is formed.

Salt Formation with Dimethyl Pyrone and the Tetravalency of Oxygen

Dimethyl – pyrone (i) has been usually be Collie and Cickle as the basic of an investigation into the tetravalecy of oxygen. It may be prepared by condensing the copper salt of aceto – acetic ester. With phosgene and boiling the product so obtained with sulphuric acid.

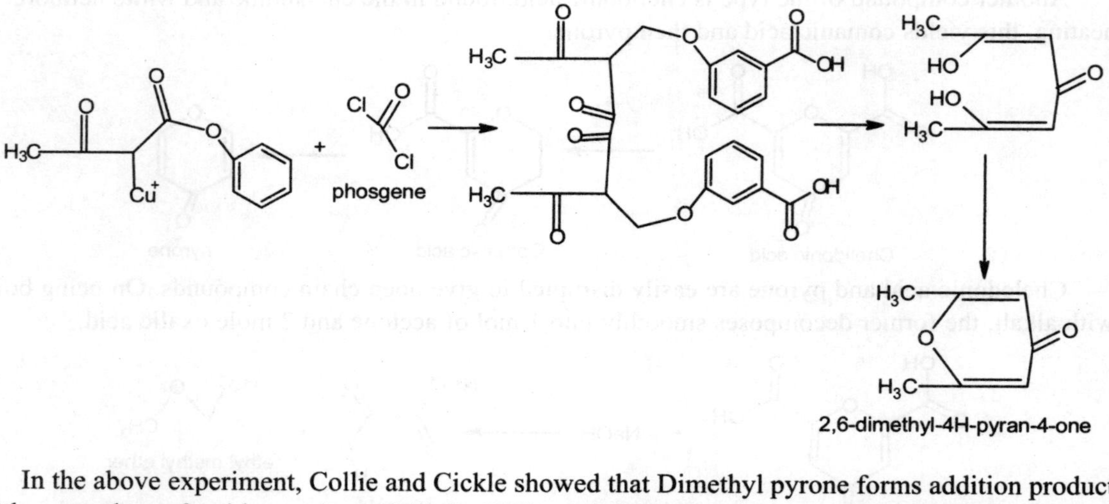

phosgene

2,6-dimethyl-4H-pyran-4-one

In the above experiment, Collie and Cickle showed that Dimethyl pyrone forms addition products with a number of acids, such as $C_7H_8O_2$ HCl with hydrochloric acid $(C_7H_8O_2)_2H_2PTCl_6$ with hydrochloroplatinic acid, and $(C_7H_8O_2)_2\ C_4H_6O_6$ with tartaric acid. It will be seen that those all result from the direct addition of acid, with out loss of water.

The stability and behaviour of these compounds can be explained on the assumption of a tetravalent oxygen atom with basic properties. It is therefore appears that oxygen can take place of sulphur, phosphorous and nitrogen hypothetical base which is known as oxonium hydroxide, H_3OOH, by analogy with the hypothetical bases, NH_4OH PH_4OH H_3SOH H_2IOH. Salt of this oxygen bases are called oxonium salts.

In this connection it will be seen that dimethyl – pyrone contains two oxygen atoms, leading to the possibility of either of the formulæ Ia or Ib

Ia

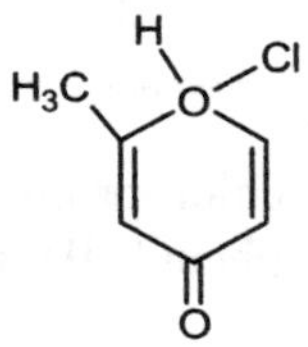

Ib

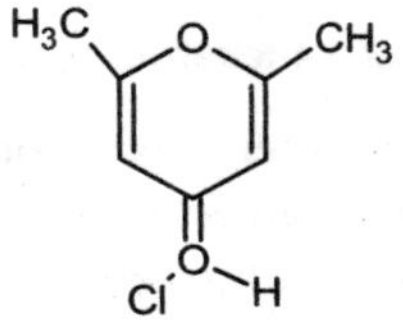

for dimethyl – proton salts. So far it has been found possible to describe with, although the carbonyl oxygen possesses more free affinity than the oxygen of ethers.

An examination of the electrical conductivity of dimethyl – pyrone salts in aqueous solution indicates that they are almost completely hydrolyzed. Dimethyl pyrone also unites with methyl sulphate to give and addition products, which on treatment with potassium iodide yields dimethyl – pyrone methiodide, of the composition CH_3I dimethyl pyrone.

Salts are also formed by the parent compound pyrone (e.g. hydrochloride picrate and oxalate), but in this case there is a strong tendency towards the formation of more complex salts. Pyrone combines, in addition, with inorganic salts such as calcium chloride, mercuric chloride and silver nitrate. In this respect is resembles the amino acids, in which Strecter assumes that the carbonyl group binds the metallic radical, and is in agreement with Walden's assumption based on conductivity experiment, that dimethyl – pyrone is an amphoteric electrolyte.

The discovery of the salts of dimethyl pyrone and their formulation as oxonium salts , has stimulated research into the question as to whether salt formation in a property of oxygen compounds in general. Experiments results obtained by Bayer and Villiger indicate that this is the case, and in recent years evidence has been supplied by other investigators confirming the existence of a great variety of addition compounds, which are regarded as salts of tetravalent oxygen.

BENZO AND DI BENZO γ PYRONE

The cyclic oxide chromane (formula III) below) may be regarded as the parent compound of a number of derivatives such as the chromones and coumarins containing the atomic network (I) Chromone has been

I II III

prepared from the base tetrahydro – quinoline, this is transferred into chromane through some intermediate reactive products.

1,2,3,4-tetrahydroquinoline → chromane

The nitrogen ring of the latter may be opened to give o - γ - chloropropyl aniline, which by means of the diazo reactions can be converted into o - γ chloropropyl – phenol (II). In alkaline solution this is quantitatively transformed into chromane.

Chromane is very strong refractive liquid which smells like peppermint and boils at 214°C. It dissolves in concentrated sulphuric acid giving a pink solution.

Physcial Properties

Molecular Formula	= $C_9H_{10}O$
Formula Weight	= 134.1751
Composition	= C(80.56%) H(7.51%) O(11.92%)
Molar Refractivity	= 40.03 cm^3
Molar Volume	= 127.3 cm^3
Parachor	= 318.1cm^3
Index of Refraction	= 1.541
Surface Tension	= 38.9 dyne/cm
Density	= 1.053 g/cm^3
Polarizability	= 15.87 $10^{-24}cm^3$
Monoisotopic Mass	= 134.073165 Da
Nominal Mass	= 134 Da
Average Mass	= 134.1751 Da
Log P	= 2.89

From benzo pyrone (chromone and dibenzopyrone xanthone) are derived a number of naturally occurring yellow dyes, the colour of which due to the chromophore CO.

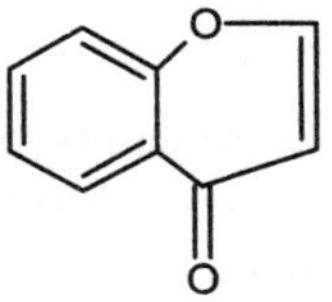

Chromone 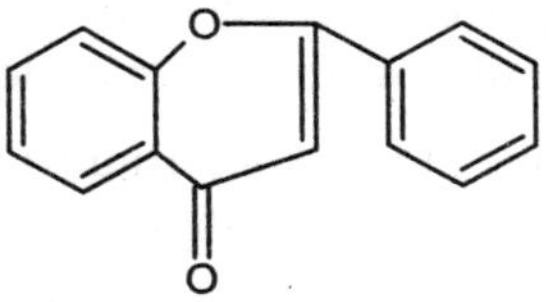   Flavone 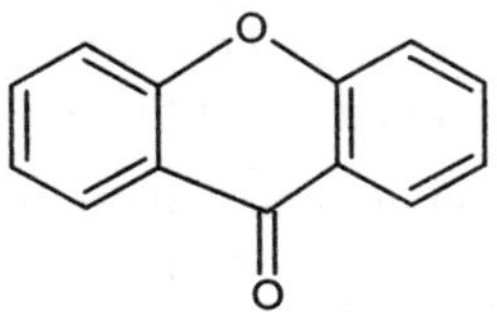   9*H*-xanthen-9-one

Chromone (or **1,4-benzopyrone**) is a derivative of benzopyran with a substituted keto group on the pyran ring.

Flavones are a class of flavonoids based on the backbone of 2-phenylchromen-4-one (2-phenyl-1-benzopyran-4-one) shown on the right. Natural flavones include Apigenin (4',5,7-trihydroxyflavone), Luteolin (3',4',5,7-tetrahydroxyflavone) and Tangeritin (4',5,6,7,8-pentamethoxyflavone). Synthetic flavones are Diosmin and Flavoxate.

Luteolin is a flavonoid and more specifically a flavone. It is thought to play an important role in the human body as an antioxidant, a free radical scavenger, an agent in the prevention of inflammation, a promoter of carbohydrate metabolism, and an immune system modulator. These characteristics of luteolin are also believed to play an important part in the prevention of cancer. Multiple research experiments describe luteolin as a biochemical agent that can dramatically reduce inflammation and the symptoms of septic shock.

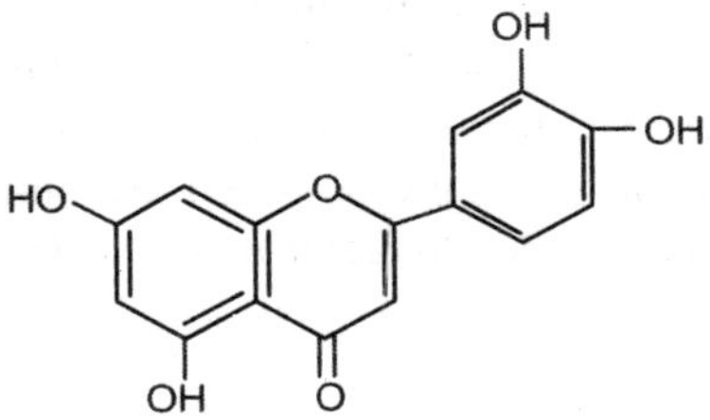

OH OH HO OH O

Tangeritin is a polymethoxylated flavone that is found in tangerine and other citrus peels.

Although few randomized, double-blind human studies have been done, animal research shows the potential of tangeritin as a cholesterol lowering agent. A hamster study showed potential protective effects against Parkinson's disease. Tangeritin shows potential as an anti cancer agent. In in vitro studies, tangeritin appears to counteract some of the adaptations of cancer cells. Tangeritin strengthens the cell wall and protects it from invasion. Tangeritin induced apoptosis in leukemia cells while sparing normal cells [3]. It counteracts tumor suppression of gap junction intercellular signaling. It acts to freeze cancer cells in phase G1 of the cell cycle, preventing replication. In summary, *in vitro* studies show antimutagenic, antiinvasive and antiproliferative effects. One caveat is that tangeritin appears to counteract the anticancer drug tamoxifen and to suppress the activity of natural killer cells. Tangeritin is commercially available as a dietary supplement.

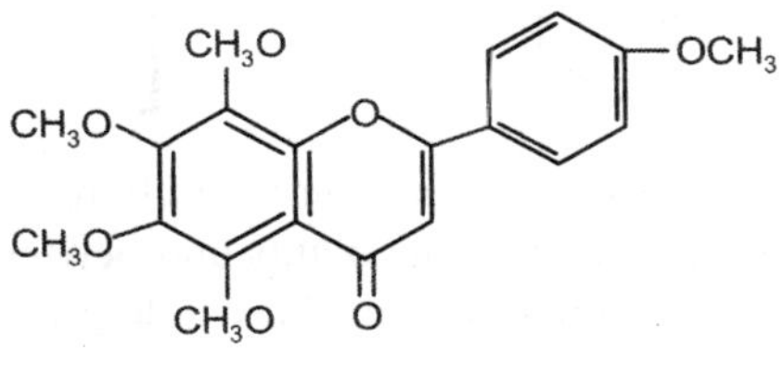

CH_3O OCH_3 CH_3O CH_3O CH_3O O

Diosmin is a semisynthetic phlebotropic drug, a member of the flavonoid family. It is used with Hesperidin to control internal symptoms of hemorrhoids (piles). It is an oral phlebotropic drug used in the treatment of venous disease, i.e., chronic venous insufficiency (CVI) and hemorrhoidal disease (HD), in acute or chronic hemorrhoids, in place of rubber-band ligation, in combination with fiber supplement, or as an adjuvant therapy to hemorrhoidectomy, in order to reduce secondary bleeding.

Xanthone is an organic compound with the molecular formula $C_{13}H_8O_2$. It can be prepared by the heating of phenyl salicylate.[2] In 1939, xanthone was introduced as an insecticide. Xanthone currently finds uses as ovicide for codling moth eggs and as a larvicide.[3] It is also used in the preparation of xanthydrol, used in the determination of urea levels in the blood. Xanthones are natural constituents of plants in the families Bonnetiaceae and Clusiaceae. Xanthones are also reported from some species in the family Podostemaceae.

XANTHYLIUM AND PYRYLIUM SALTS

Xanthylum salts stand in close relationship with triaryl methylium salts (phenanthraquinone). They only differ from the latter in containing an oxygen bridge between two of the benzene nuclei, and hence the two classes of compounds have very similar properties. Goumberg and Cone showed that the xanthylium compounds, although colourless, yield coloured products by further addition of acids or salts, The xanthylium formed with oxygen acids, however, are without exception coloured. In these respects there is a complete resemblance to the triaryl methane derivatives; Kehrmann found that in certain special cases even the simple halogenides of the xanthyl series may be coloured and then also possess the character of salts. By analogy with the triaryl compounds as in II.

(I) (II)

O X O+ X H_3C O CH_3 ClO_4 CH_3

The xanthylium salts are closely related to the benzapyrylium salts, occurring in the anthocyanidin colouring matter of the plants and berries and to the simple pyryium salts, which may be regarded as the parent compounds of the whole group.

PYRIDINE GROUP

Pyridine and its derivatives contain a ring composed of five carbon atoms and one nitrogen atom. Pyridine can therefore be derived from benzene by replacing a trivalent group (CH) by an atom of nitrogen, its parent compound of are number of vegetable alkaloids.

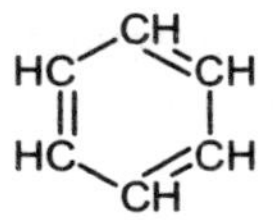

benzene 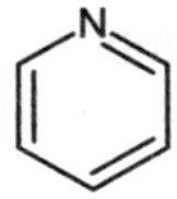 pyridine

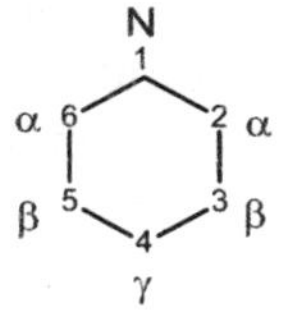

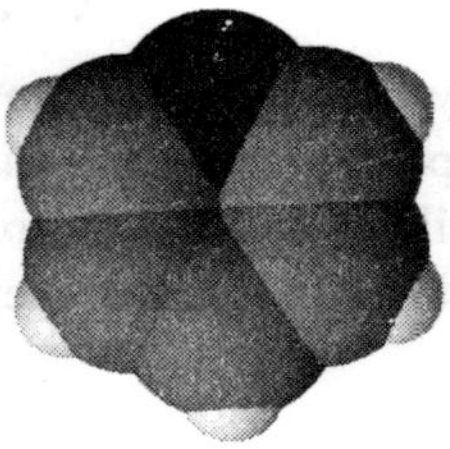

The above formula was proposed by Körner in 1869, offers a satisfactory explanation of the chemical behaviour of pyridine and its derivatives and of the well marked analogy between benzene and it has been confirmed by several synthesis of pyridine compounds. As in the case of benzene other formulæ have also been put forward.

The possibilities of isomerism among derivatives of pyridine are greater than with benzene, since not only does the relative position of the constituents to one another enter into the question, but also their position, with regard to the nitrogen of the ring. Isomerides are usually described by the use of numbers or letter as indicated in the above formulæ. Theory predicts the existence of three mono substitution products, and six or twelve distribution according as the substituents similar or dissimilar.

Preparation

Many methods exist in industry and in the laboratory (some of them named reactions) for the synthesis of pyridine and its derivatives:[3] Pyridine was originally isolated industrially from crude coal tar. It is currently synthesized from acetaldehyde, formaldehyde and ammonia, a process that involves the intermediacy of acrolein:

$$CH_2O + NH_3 + 2\,CH_3CHO \rightarrow C_5H_5N + 3\,H_2O$$

By substituting other aldehydes for acetaldehyde, one obtains alkyl and aryl substituted pyridines. 26,000 tons were produced worldwide in 1989.[4]

- The Hantzsch pyridine synthesis is a multicomponent reaction involving formaldehyde, a keto-ester and a nitrogen donor.
- Other examples of the pyridine class can be formed by the reaction of 1,5-diketones with ammonium acetate in acetic acid followed by oxidation. This reaction is called the "Kröhnke pyridine synthesis."
- Pyridium salts can be obtained in the Zincke reaction.
- The "Ciamician-Dennstedt Rearrangement" (1881) is the ring-expansion of pyrrole with dichlorocarbene to 3-chloropyridine and HCl[5]
- In the "Chichibabin pyridine synthesis" (Aleksei Chichibabin, 1906) the reactants are three equivalents of a linear aldehyde and ammonia

The simplest method of building up a pyridine ring is from aliphatic compounds of the genral formula.

$$X—CH_2—CH_2—CH_2—CH_2—CH_2—NH_2$$

Which may be converted into piperidine by ring formation, and by subsequent oxidation yield pyridine. Thus pentamethylene – diamine hydrochloride on decomposes, in to ammonium chloride and piperidine. Similarly normal chloro – and ω - bromo – amylamine lose hydrogen halide on heating with alkali, to give piperidine (Landenberg).

$H_2C(CH_2—CH_2—NH_2\,HCl)(CH_2—CH_2—NH_2HCl)$ —(-NH_4Cl)→ piperidine (H_2C, CH_2—CH_2, NH, CH_2—CH_2)

$H_2C(CH_2—CH_2—Br)(CH_2—CH_2—NH_2)$ —(-HBr)→ piperidine

piperidine —oxidation→ pyridine

2. A pyrogenic synthesis of pyridine, analogous to that of benzene from acetylene occurs when acetylene and hydrogen cyanide are lead through a tube, which is heated to redness.

$$2C_2H_2 + HCN \rightarrow C_5H_5N$$

Physical Properties

Molecular Formula	= C_5H_5N
Formula Weight	= 79.0999
Melting Point	= -41.6 °C
Boiling point	= 115.2 °C
Composition	= C(75.92%) H(6.37%) N(17.71%)
Molar Refractivity	= 24.34 cm^3
Molar Volume	= 82.6 cm^3
Parachor	= 201.4 cm^3
Index of Refraction	= 1.500
Surface Tension	= 35.2 dyne/cm
Density	= 0.956 g/cm^3
Polarizability	= 9.65 cm^3
Monoisotopic Mass	= 79.042199 Da
Nominal Mass	= 79 Da
Average Mass	= 79.0999 Da
Log P	= 0.73

Chemical Properties

Pyridine is a colourless liquid of unpleasant penetrating smell. It is miscible with all proportion in water, alcohol and ether and forms salts with acid. Among the latter ferrocyanide and perchlorate of potassium are difficultly soluble and are used for the isolation and purification of pyridine.

The outstanding feature of pyridine is its great stability. Chromic acid potassium permanganate and nitric acid cannot react with it, and with sulphuric acid it is converted into sulphonic acid derivative at about 300°C. halogen – substitution products are only obtained with great difficulty, the halogen entering β - position.

Mercuration, on the other hand proceeds readily. Pyridine is a tertiary base, and when reduced with sodium and alcohol yields the secondary base piperidine.

Pyridine sulphuric anhydride is prepared as crystalline compound by drawing carefully regulated current of air impregnated with sulphur trioxide (from warm oleum) and pyridine, into a barrel – shaped receiver. The product used for the preparation of sulphuric esters of phenolic compounds, in place of the mixture of chloro sulphonic acid and pyridine previously employed. It is much less sensitive to moisture than the usual sulphonating agents, and as a solid is more easily handled and weighted. The addition compound from pyridine and sodium pyrosulphate may be used in a similar manner.

Formation of Derivatives of Pyridine

The formation of pyridine derivatives take place from pyrrole, by extension of the ring, which has been discussed previously.

Pyridine derivatives can also be prepared from compounds of the pyrone group by treatment with ammonia, when the oxygen atom of the ring is replaced by the NH group.

The parent compound and its homologous are tertiary bases, which unite with one equivalent of an acid to form salts and also combine with inorganic salts such as mercuric chloride, and the sulphate of copper, zinc and cadmium to give double salts of type $C_5H_5N.HgCl_2$ and $(C_5H_5N)_2$ $(HgCl_2)_3$ as shown clearly before the composition and behaviour of these addition compounds illustrate the similarity between the pyridine and pyrazole series.

Pyridine and its derivatives unite directly with sodium bisulphite. The compounds so formed readily decompose with loss of ammonia and simultaneous opening of the ring.

The strong resemblance between the pyridine and benzene serried is emphasized by the following facts. Oxidizing agents such as nitric acid and chromic acid attract neither benzene or pyridine. Potassium permanganate converts pyridine homologues, the side chain being oxidized while the constitution of the pyridine – carboxylic acids so obtained, conclusions can be drawn to the number and position of the side chains originally present. Sulphuric acid converts its homologues into sulphonic acids although the action is slower than benzene. The sulphonic group in these acids can be exchanged fro the hydroxyl or cyano – group by fusion with potassium hydroxide or cyanide respectively. The resulting hydroxy pyridines resembles the phenols in behaviour.

One notable difference between pyridine ring can only be nitrated with great difficulty (360°C) unless it is first substituted by some group such as OH or NH_3. According to Marckwald the resistance of pyridine to nitration is due to the strongly negative character of the nitrogen atom. Thus electrophilic substituents (NO_2, CL, Br SO_3H) always enter the β - position in the ring, i.e. the meta position to the nitrogen. The negative character of the nitrogen atom is also revealed in the α – γ – chloro – compounds of pyridine, the halogen of which is mobile as in the o and p – chloroderivatives of nitrobenzene. The reactivity of the halogen in α and γ chloropyridines is shown by the conversion of these compounds into amino – pyridines with ammonia pyridyl – hydrazines with hydrazines and mercaptans with potassium hydrosulphide.

Pyridine and its homologues react with sodamide to form α - aminopyridines. In this case the unusual point of attack is due to the nucleophilic character of the entering substituent ($-NH_2$).

The α and γ — methyl — pyrilidine are also unusually reactive. According to experimental conditions they either condense with aldehydes to form products of the aldol type known as alkines e.g. compound I or else water is eliminated and oxygen free, unsaturated bases, such as α - allyl pyridine (II) are produced. The latter are generally termed as stilbazoles. As will be seen later, the α - allyl pyridine obtained by this reaction is an intermediate product of the synthesis of the alkaloid conine.

Phthalic anhydride and phalimide may also be employed in place of aldehydes in this condensation.

(I) $CH_2-HC(OH)-CH_3$

acetaldehyde

(II) $CH=CH-CH_3$

On reducing pyridine bases with sodium and alcohol, six atoms if hydrogen are taken up with formation of piperidine bases. More energetic reduction, by heating with hydrogen iodide, ruptures the ring with the production of paraffins, e.g. pyridine is converted in pentone.

Homologues of Pyridine

The alkyl derivatives of pyridine mentioned above are found together with pyridine itself in bone oil and coal tar.

Hydroxy and Amino Pyridines

Hydroxy – pyridines may be compaired with amino – phenols which they resemble in yielding salts with bases as well as with acids.

The three hydroxy – pyridines are high boiling substance which crystalline well they are best prepared from the corresponding carboxylic acids by elimination of carbon dioxide, and are also formed by direct hydroxylation when pyridine vapour is led over powdered potassium hydroxide qt 300°C to 320°C. A point of special interest is the tautomerism exhibited as true hydroxy pyridine, containing a phenolic hydroxyl group as pyridine or ketonic derivative of a dihydro pyridine.

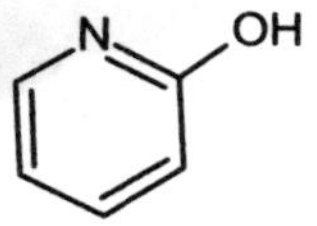

α - Hydroxypyridine 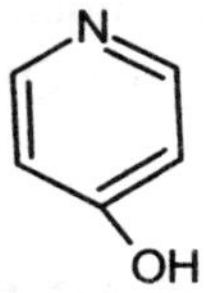   pyridin-4-ol 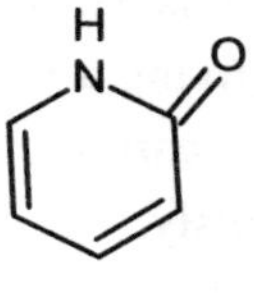   pyridin-2(1*H*)-one 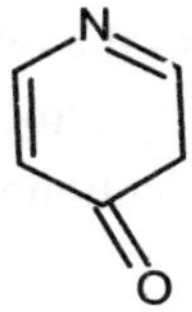   pyridin-4(3*H*)-one

Where as the free α and γ - hydroxy pyridine have also far only been isolated in one form and it is sill uncertain whether this corresponds to the compounds each exist in two forms of the constitution $C_6H_4(OR)N$ and $C_6H_4O(NR)$ respectively, β - hydroxy pyridine on the other hand, reacts only as a phenol and never according to the pyridone type.

In di and tri hydroxy pyridines the basic character is entirely lost. These also show tautomerism of the above kind

It has been stated that pyridine reacts with sodamide to form α - .amino and αα' – diamine pyridines. BY analogy with the case of a hydroxy pyridine, it would be expected that α - hydroxy – pyridine, it would give rise to two series of derivatives, corresponding to the tautomeric forms I and II, which may ge described respectively as α - amino pyridine and a pyridone – imide.

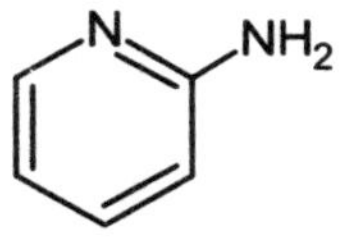

pyridin-2-amine
α - amino pyridine

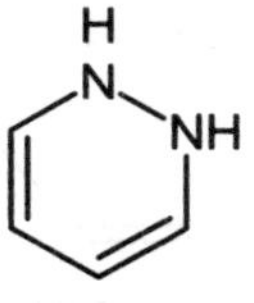

1,2-dihydropyridazine

In actual practice alkyl derivatives of both tautomeric forms are easily prepared. A valuable antiseptic, pyridium β - phenylazo – 2: 6 – diamine pyridine.

PYRIDINE CARBOXYLIC ACID

Carboxylic acids of the pyridine series result, as stated above, from the oxidation of pyridine derivatives containing organic side chains. Hence, they are frequently obtained as degradation products of vegetable alkaloids, and knowledge of their constitution is of great value in investigating the structure of the later. They are solid compounds, prossessing both acidic and alkaline character, although the basic properties are not very evident in poly – carbolic acids. When heated with lime all these acids decompose into pyridine and carbon dioxide. A carbonyl group in the α position is particularly readily removed; a acids also differ from the others in giving a yellowish – red colouration with ferrous salts. Pyridine mono – carboxylic acids $C_5H_4(COOH)N$ have already been mentioned previously.

HYDRO – PYRIDINE DERIVATIVE

Derivatives of Dihydro – pyridine are produced synthetically as stated before by the action of ammonia on diketo compounds.

On reducing pyridine and its derivative by means of sodium and alcohol, six atoms of hydrogen are taken up and hexahydro – compounds produced, 1:4 dihydropyridine being formed as n inter mediate products.

The reduction of pyridine with zinc dust and acetic acid anhydride leads to the formation of N, N diacetal – (tetra) hydro γγ' – dipyridyl) of formula II below, this is a crystalline compound, (m.p 124°C), the production of which is explained by assuming that the reagents first attack the nitrogen atom with the temporary formation of the radical (I), which by union with itself yields the dipyridal derivative.

(I)

1-acetyl-1,4-dihydropyridine

(II)

1,1'-diacetyl-1,1',4,4'-tetrahydro-4,4'-bipyridine

PIPERIDINE

Introduction

Piperidine is an organic compound with the molecular formula $C_5H_{11}N$. It is a heterocyclic amine with a six-membered ring containing five carbon atoms and one nitrogen atom. It is a clear liquid with a pepper-like odor. The piperidine structural motif is present in numerous natural alkaloids such as piperine and quinine, and is the main active chemical agent in black pepper and relatives (Piper *sp.*), hence the name. Piperidine is also a structural element of many pharmaceutical drugs such as raloxifene, minoxidil, thioridazine and mesoridazine. Piperidine is often used as a solvent for its mild basic properties, most notably in Fmoc-strategy synthesis. The major industrial application of piperidine is for the production of dipiperidinyl dithium tetrasulphide, which is used as a rubber vulcanization accelerator.

Piperidine is naturally found in fire ant venom, and is the cause of the burning sensation associated with the bite of these insects. Piperidine is also commonly used in chemical degradation reactions, such as the DNA sequencing method invented by Walter Gilbert in 1977, for cleavage of particular modified nucleotides. Piperidine is also commonly used as a strong base for the deprotection of amino acids in solid-phase peptide synthesis.

Preparation

This was first prepared, by heating alkaloid piperine with alkali; it is also prepared by the reduction of pyridine (with sodium and alcohol or by electrolytic means) has already been mentioned, It is a colourless liquid of peculiar Ammonical smell, miscible in all proportions with water, alcohol and benzene.

Physcial Properties

Molecular Formula	$= C_5H_{11}N$
Formula Weight	= 85.14754
Composition	= C(70.53%) H(13.02%) N(16.45%)
Melting point	= –7 °C
Boiling point	= 106 °C
Molar Refractivity	= 26.40 cm^3
Molar Volume	= 102.4 cm^3
Parachor	= 233.0 cm^3

Index of Refraction	= 1.428
Surface Tension	= 26.7 dyne/cm
Density	= 0.830 g/cm^3
Polarizability	= 10.46 cm^3
Monoisotopic Mass	= 85.089149 Da
Nominal Mass	= 85 Da
Average Mass	= 85.1475 Da
Log P	= 0.93

Chemical Properties

The imino hydrogen atom of the piperidine may be replaced by different radicals (alkyl, acyl, nitrose groups, etc) and numerous derivatives have thus been prepared which cannot be described here. On being heated at 300°C with concentrated sulphuric acid at 250°C with nitrobenzene are at 180°C with sliver acetate piperidine becomes oxidized, to pyridine.

Alkaloids derived from piperidine are treated in the chapter 36. But it has been shown by J. V Braun, alkylation of the carbon atom strengths the structure of the piperidine ring. This effect becomes apparent on the introduction of a single methyl group.

Methods of Opening the Piperidine Ring

A number of methods are available for rupturing the piperidine ring processes which are in a sense a reversal of the synthesis described under the sub head of the synthesis of piperidine

1. By oxidation: —Under the influence of oxidizing agents, such as hydrogen peroxide, the piperidine ring is comparitavely easily broken between the nitrogen atom and an adjacent carbon atom, with formation of δ - amino – valeraldehyde.

piperidine —Oxygen→ glutaraldehyde

The opening of the ring is effected even more readily by the action of potassium permanganate on N – acylated piperidine derivatives> in this way benzoyl – piperidine, $C_5H_{10}N(COC_5H_5)$ yields benzoyl δ amino valeric acid. On the other hand, when N – Alkyl piperidines are treated with hydrogen peroxide, they take up an oxygen atom to form alkyl piperidine oxides.

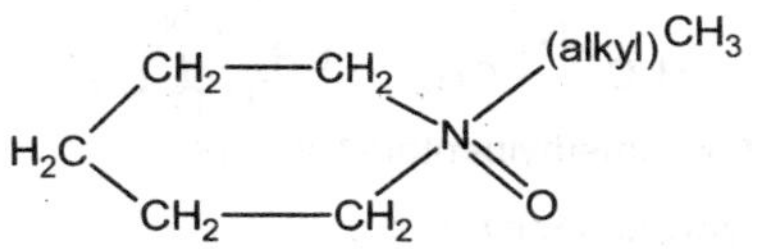

2. *By means of phosphorous Halides*: The work of J, V Braun has shown that acyl derivatives of piperidine are very easily attached by phosphorous pentachloride or pentabromide. Under chosen conditions the resulting 1: 5 – dichloro – pentane or 1:5 dibromo pentane is obtained in so good a yield that the reaction can be used as a means of preparing these halogen compounds.

3. *By means of Cyanogen Bromide*: The substitution products of piperidine and other cyclic tertiary bases of the general type X 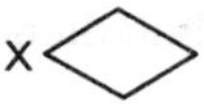 N.R are disrupted by cyanogen bromide according to the following equation to give brominated cyanaides Br.XN(CN)R.

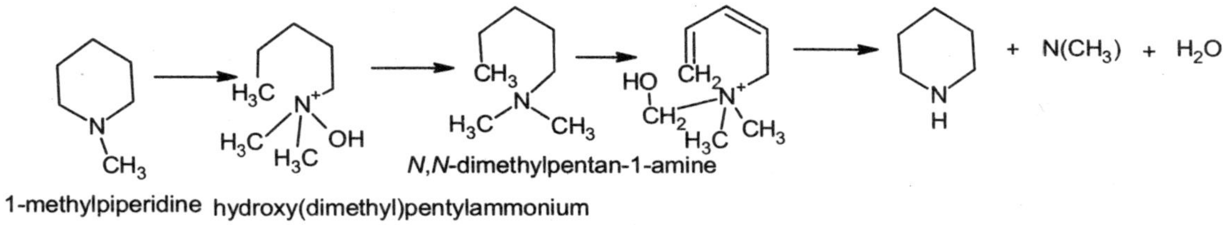

+ N(CH_3) + H_2O

1-methylpiperidine hydroxy(dimethyl)pentylammonium N,N-dimethylpentan-1-amine

Provided that the alkyl group $CH_3 - CH_2$ is not removed from the molecule. Such a cynamide derivative may under go hydrolysis to a brominated secondary amine Br.CH_3N(CN) $CH_3 - CH_2$. This constitutes the simplest available method of opening ring containing nitrogen, and may also be employed with success in cases where (as with the aromatic derivative of piuperidine) the following method (4) cannot be applied.

4. *By means of Exclusive Methylation*: A method of opening the piperidine ring, with simultaneous loss of nitrogen is by "exclusive methylation" This series of reactions, first used by A.W Hoffmann in the case of piperidine was correctly interpreted by Ladenberg and later applied by other investigators to a large number of cyclic bases. It has been the classical weapon to attack in determining the constitution of the majority of vegetable alkaloids, and may therefore be treated in some detail. The operations are also as follows: Piperidine as a secondary base, can be methylated at the nitrogen atom by means of methyl iodide.

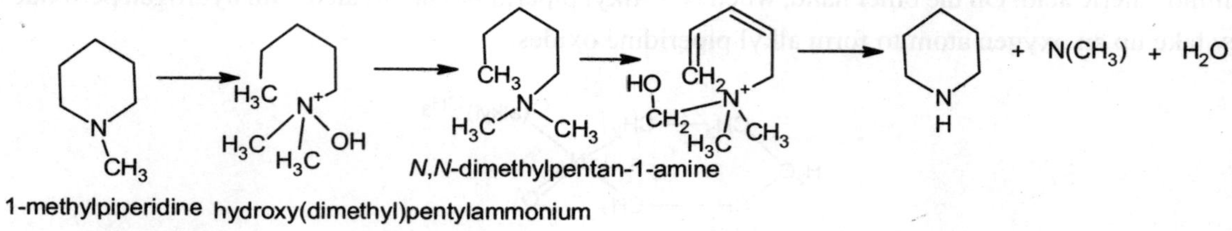

+ N(CH$_3$) + H$_2$O

N,N-dimethylpentan-1-amine

1-methylpiperidine hydroxy(dimethyl)pentylammonium

The methyl piperidine so obtained unites with methyl iodide to from dimethyl piperidium iodide, and this by treatment with moist potassium hydroxide is converted into dimethyl piperidinium hydroxide. The latter on distillation breaks up into water and a compound frequently describes as dimethyl – piperidine, but currently named Δ^4 phenyl – dimethylamine. Being a tertiary base this substance also unites with methyl iodide to form a substituted ammonium hydroxide and submitted to dry distillation yields trimethyl amine, water and hydrocarbon α - methyl butadiene of the formula C_6H_8.

CHAPTER

36

The Alkaloids

Introduction

Alkaloids are nitrogenous compounds having basic properties, usually combines with acids, without elimination the water, as it behaves almost like alkalies, the name comes from its function.

Many of these compounds possess curative properties and are of great value in medicine. Although poisonous and therapeutic properties of various plants have been known as utilized from early times, it was not until 1817 that the first alkaloid was isolated. A large number of these now known but for a long time all attempts to determine their constitutions or to prepare them were fruitless.

The chemistry of alkaloids being to make definite progress with the discovery of pyridine and quinoline, which led to the view that they were related to these base in the same manner as aromatic compounds to the benzene, Röuing's in 1880, defined alkaloids as derivatives of pyridine, or indeed as belonging to any single class of organic compound, had to be obtained, owing to the discovery that natural groups of vegetable bases, such as the morphine and coca groups cannot be referred to any one parent substance but belong to a number of different systems.

In describing all these vegetable bases as alkaloids, we are therefore collecting into one class a number of substances of widely differing constitutions. A few of these contain nitrogen in an open chain but in isoquinoline or pyrrole. Still other alkaloids are derives from purine or from complex dicyclic systems such as are contained in the "second half" of the cinchona alkaloids and in the tropine group.

Preparation of Alkaloids from Plants and General Properties

Alkaloids are usually found in the plants in the form of salts – in which they are either united to the common plant acids (e.g. malic or citric acid) or to certain characteristic acids such as quinic acid, in the cinchona alkaloids and mecanoic acid) in those of the opium group. Their distribution is very unequal. Although they may be detected in all parts of the plant, they generally accumulate in the fruit and seeds, and also in the bark of the trees.

In preparation of alkaloids from plants the finely divided material is usually extracted with water containing hydrochloric or sulphuric acid. This liberates the alkaloids from their salts with organic acids and the bases pass into solutions as hydrochlorides or sulphates, together with the dye stuffs, carbohydrates and other products from the plant tissue. From the solution so obtained the alkaloids, being insoluble or only sparingly soluble in water, can be prepared by the addition of alkali. If the bases are volatile, as in the case of nicotine, the solution or finely divided raw material is treated with alkali and distilled in steam. The crude alkaloids are then purified by special methods, frequently by recrystallization of the free compounds or their salts.

The majority of the alkaloids are solid substances which cannot be distilled only a few, such as coniine are liquid and volatilize without decomposition. On the animal organism as has already been mentioned they often exert a marked physiological action.

Almost without expectation they are either insoluble or sparingly soluble in water, dissolve with difficulty in chloroform, either and benzene, and readily in alcohol. Most of the alkaloids are optically active and usually lævorotatory. In many cases their solutions gives a strong alkaline reaction.

All of them from salts with acids, among which the hydrochlorides sulphates and oxalates crystallize particularly, possess the property of uniting with certain metallic salts such as the chlorides of mercury, platinum and gold, to form double compounds.

Alkaloids are precipitated from the aqueous or acid solution by a number of substances generally known as '*alkaloid agents*' e.g. tannic acid, picric acid, picrobromic acid, perchloric acid, potassium bismuth iodide, phosphomolybdic acid and phosphotungstic acid. These reagents however are of no great value for the quantitative analysis of alkaloids, since the resulting compounds are not sufficiently insoluble and because the reagents also precipitate other organic substance.

Methods of Determining the Chemical Composition of Alkaloids

In attempting to determine the structure of an alkaloid on the first task is to investigate the action of '*hydrolyzing agents*'. When heated with water, acids or alkalies many of the vegetable bases break up into a characteristic alkaloidal constituent, containing nitrogen, and a nitrogen free component. In general, the latter consists of an acid, the carboxyl group or with an alcoholic hydroxyl of the nitrogenous constituent; in the comparatively rare gluco – alkaloids, among which is numbered solanine; second component is a sugar.

Thus piperine decomposes on hydrolysis into piperidine and piperic acid; the union in this case is of the acid amide type,

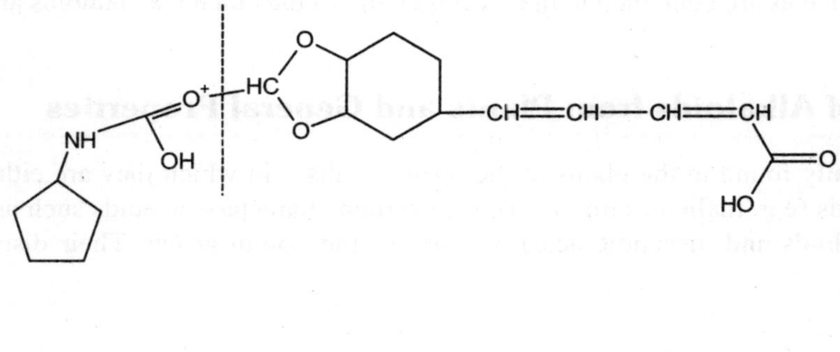

O⁺ HC O O NH OH CH═CH—CH═CH O HO

Atropine, as well be seen latter may be hydrolyzed to give tropic acid and the alkaline tropine.

A second method is to effect degradation by distillation with zinc dust fusion with alkali, or heating with bromine or other vigorous reagents, as a result of which some stable parent substances can often be isolated.

Gerhardt, as early as early as 1842, obtained quinoline from cinchonine by distilling the latter with alkali, Vongerichten and Schröter isolated phenanthrene as the main product of the distillation of morphine with zinc dust. Alkaloids containing oxygen generally loose this element on treatment with zinc dust, while those rich in hydrogen become dehydrogenised. Our knowledge of the constitution of coniine, for example, is based in Hoffmann's observation that on distilling the compound with zinc dust it looses six hydrogen atoms to give conyrine (2 – propyl – pyridine)

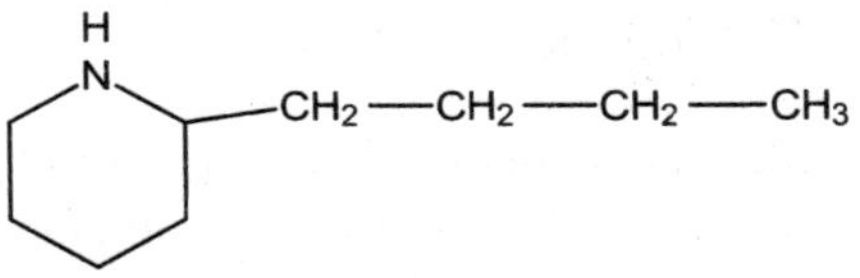

$-CH_2-CH_2-CH_2-CH_3$

2-butylpiperidine

$-CH_2-CH_2-CH_3$

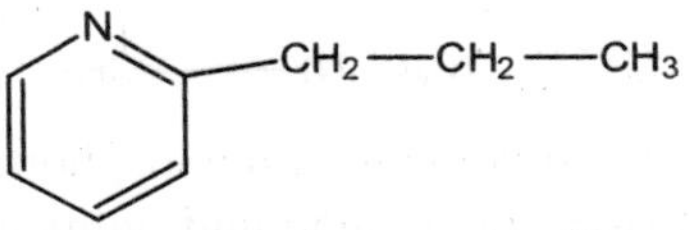

2-propylpyridine

Other methods of remaining hydrogen have already been quoted under piperidine. Meroquinene has been shown to be a pyridine derivative by Könings, who obtained 3-ethyl-4-methyl-pyridine by heating it with hydrochloric acid and mercuric chloride.

In the hydrogenation of alkaloids an important part is played by '*catalytic method*' of reduction, involving the use of metals of the platinum group. In this manner morphine readily yields dihydro – morphine, and the cinchona alkaloids, give dihydro derivatives.

In addition, the estimation of the methyl imino group, as developed by Herzig and Meyer is used in the alkaloid investigation.

When the hydriodides of N-methylate bases are heated at 200°C to 300°C, they part with methyl iodide according to the equation.

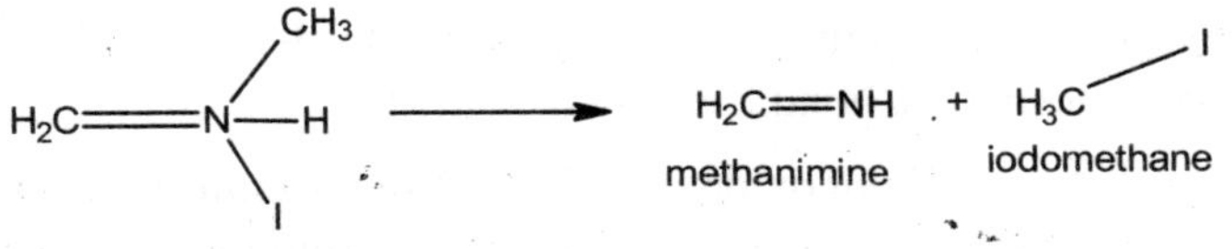

$$H_2C{=}N(CH_3)(H)I \longrightarrow H_2C{=}NH + H_3C{-}I$$

iodo(methyl)methylene-λ^5-azane methanimine iodomethane

The methyl iodide can be estimated by Zeisel's method in which it is absorbed in an alcoholic silver nitrate solution weighted.

Zeisel's method for the determination of methoxy group depends upon the conversion of the methyl of the CH_3O – group into methyl iodide, by the treatment with hydrochloric acid at the boiling point and the subsequent estimation of iodide in the above manner. A process for the determination of methoxyl in the presence of methyl imino group (Herzig and Meyer) is based on the fact that the methoxyl group is hydrolysed at the boiling point of hydroiodic acid, whilst the N-methyl group is not detailed until a higher temperature has been reached, with regard to the value of this method for distinguishing between a methoxyl and a negative result at the lower temperature may safely be taken to prove the absence of methoxyl, but otherwise no certain conclusion can be drawn without further information.

Fourthly, *the function of the oxygen atoms in the alkaloid* must be investigated, in this connection a reaction of special importance is the conversion of an alkaloid containing as alcoholic hydroxyl group into its anhydro-compound be means of dehydrating agents such as a solution of glacial acetic acid in sulphuric acid (e.g. tropine "tropidine").

$$(C_6H_{11}N)\text{—}CH_2\text{—}CH\text{—}OH \longrightarrow (C_6H_{11}N)\text{—}CH{=}CH\text{—} + H_2O$$

Or by successive treatment with phosphorus chlorides and alcoholic potash (e.g. production of cichene and quinine) from cinchonine and quinine). The unsaturated compounds so obtained are often more reactive than the original alkaloids, and can with advantage be substituted to reactions involving further degradation.

Alcoholic and phenolic hydroxyl group are estimated in the usual manner by acceptation and benzoylation.

Finally, the determination of the structure of an alkaloid generally necessities *an investigation of the oxidation products*. Towards oxidizing agents alkaloids offer a number of points of attack, such as

ethylene linkings. $>C{=}C<$, carbinol groups, $>C(H)OH$, methyl groups $>N.CH_3$ and others.

Among the reagents used for this purpose the most important are potassium permanganate, chromic acid, nitric acid and hydrogen peroxide. Permanganate is a particular value in attaching double bond between carbon atoms, when two hydroxyl groups are first added. The resulting glycols are best further oxidized by means of chromic acid, which leads to the molecule being ruptured at the point originally occupied by the double bond.

Hydrogen peroxide brings about oxidation at the nitrogen atom and opens oxidation at the nitrogen atom and opens up the ring. In the case of the saturated compounds of an aliphatic nature permanganate has the particular property often observed in the tropine – series – of oxidizing the methyl group away from hydrogen,

A good example of the use of oxidation methods will be described later in the connection with nicotine. An interesting method frequently employed in the examination of structure of alkaloids is to study the degradation products they yield on *exhaustive methylation*, by which in its widest sence is understood the decomposition of the substituted ammonium hydroxide under the influence of heat, or of employed in the exhaustive methylation of alkaloids are well illustrated by the degradation of

N-methyl-piperidine to piperylene, a classical example discovered by A. W Hoffmann. Another simple example will be found under aporphine. In this manner the carbon framework of the alkaloid molecule is revealed in the from of unsaturated hydrocarbons.

This method of decomposition may be applied to alkaloids with all conceivable groupings in the molecule and also which of special importance to amino-acids obtained by the oxidation alkaloids.

The degradation products formed in this way include a great variety of unsaturated non-nitrogenous compounds, including hydrocarbon, ketones aldehydes and carboxylic acids. For determining the structure of alkaloids, this method is therefore of general service, since these unsaturated products of exhaustive methylation can often be converted by simple reactions such as reduction, into compounds of known constitution.

Tropionic acid, for example on *exhaustive methylation* gave a diolefin dicarboxylic acid formula $C_7H_8O_4$ and of unknown structure, on a normal dicarboxylic acid containing seven carbon atom.

Hence, it follows that the carbon skeleton in tropine and ecgonine must possess an unbranched chain of seven atoms and that these are arranged in the form of a ring, since tropionic acid is produced from tropine and ecgonine by rupture of part of the cyclic system. The application of the same principle also enabled this cycloheptane ring to be isolated intact from cocaine and atropine, in the form of its ketone suberone.

The value of exhaustive methylation followed by reduction of the resulting dehydration products is not confirmed to the information this gives concerning structure, as it is often possible to affect a synthesis of the alkaloid by applying the method in the reverse direction.

The fission of cyclic bases by means of phosphorous halides (J.V Braun) has also been applied to the determination of alkaloid structure. This treatment yields open-chain halogen compounds.

Information as to the effect of the high temperature on an alkaloid can sometimes be gained by fusion with urea. The behaviour of berberine.

Classification of the Alkaloids

Under this sub-head we are going to discuss about the classification according to their chemical constitution, especially with reference to the basic compounds from which they are derived. In most cases it is then found that alkaloids produced by one and the same plant, and therefore belonging to the same botanical group, also fall into the same chemical group, owing to the fact the compounds generated by a given plant frequently possess similar chemical constitutions.

The alkaloids are therefore divided into the following groups:

I. Hydroxy-phenyl alkyl amine and phenyl hydroxy-alkyl amine bases

II. Alkaloids of pyridine group

III. Alkaloids of pyrolidine group

IV. Alkaloids of quinoline group

V. Alkaloids of isoquinoline group

VI. Alkaloids of phenathrene group

VII. Alkaloids of purine group

Like every classification this is somewhat arbitrary. It may be objected that the alkaloids atropine and cocaine, treated under the pyridine group, also contain a pyridine nucleus, and should therefore have been included in the pyridine group, by themselves. The sixth group is termed as the phenathrene group, and under this head we will see morphine, codeine and thebaine. Each of these contain a phenanthrene nucleus, but the basic complex from which they are derived has not yet been determined with certainty.

Hydroxyl-Pehyl Alkylamine and Phenyl Hydroxy – Alkylamine Bases

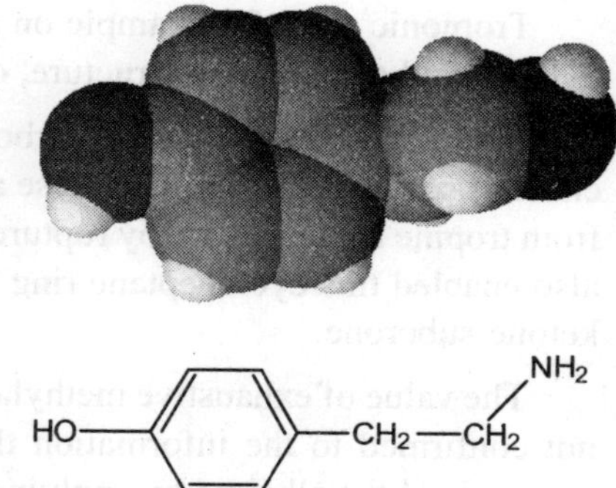

In the past years organic bases possessing phenolic character have attracted ever increasing attention, on account of their pharmacological properties. Adrenaline has become an important drug; the valuable properties of hordenine, present in fermenting barley, have been pointed out by Leger and Bayer has identified in p-hydroxy-phenyl-ethylamine one of the constituents of ergot (in diseased rai seed) which gives rise to its characteristic activity. In addition to these compounds, thyroxine, (which is described in chapter 38, the compounds of biological importance, this is synthesized by Barger to be present to the extent of 0.1 to 0.2 percent in ergot, in which it is accompanied by 4-β aminoethyl-glyoxaline (b-iminazyl- ethylamine). Physiologically, it has effect of strongly increasing the blood pressure. It may be isolated from the aqueous extract of the ergot by shaking out with amyl alcohol and crystallizes in white needles or leaflets m.p 160°C and b.p. 161°C 2mm pressure. P-hydroxy phenyl-ethylamine may be synthesized by various methods, e.g. benzyl, cyanide on reduction yields phenyl- ethylamine, the benzoyl derivative of which is converted into the hydroxy-phenyl compound by nitration, followed by reduction and diazotization. On removing the protective benzoyl group by hydrolysis, p-hydroxy phenyl-ethylamine is obtained.

NH_2

HO—⟨ ⟩—CH_2—CH_2

N-(2-phenylethyl)benzamide → N-[2-(3-nitrophenyl)ethyl]benzamide → N-[2-(2-aminophenyl)ethyl]benzamide → N-(2-phenylethyl)benzamide → 3-(2-aminoethyl)phenol

A better yield is obtained by condensing anisaldehyde with nitro methane to form p-methoxy nitrostyrole, which is then reduced and the methyl group removed with hydrochloric acid.

Anisaldehyde + nitromethane → 1-methoxy-3-[(Z)-2-nitrovinyl]benzene → 2-(3-methoxyphenyl)ethanamine → 3-(2-aminoethyl)phenol

HORDENINE

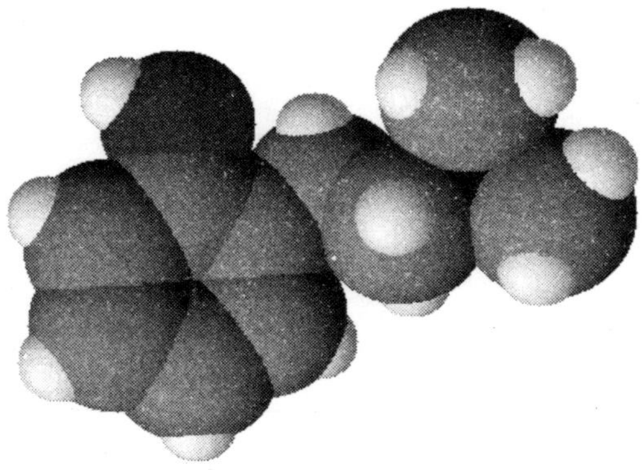

Introduction

Hordenine (*N,N*-dimethyltyramine) is a phenylethylamine alkaloid with antibacterial and antibiotic properties. It stimulates the release of norepinephrine in higher animals. It is produced in nature by several varieties of plants in the family *Cactaceae* and by some in *Acacia*.

Occurrence in Nature

Peyote (*Lophophora williamsii*), San Pedro cactus (*Trichocereus pachanoi*), and Peruvian Torch cactus (*Trichocereus peruvianus*) all produce high levels of this compound. These cacti also produce high levels of mescaline and other phenylethylamine compounds.

Cacti in the genus *Ariocarpus*, *Opuntia*, *Pereskia*, and *Coryphantha* also produce these alkaloids, though not in high concentrations. It has been shown that **hordenine**, N,N-Dimethyl-hydroxyphenylethylamine, exhibits an inhibitory action against at least 18 strains of penicillin resistant Staphylococcus bacteria.

Preparation

Hordene was first prepared by Barger from phenyl methyl alcohol.

2-phenylethanol → (2-chloroethyl)benzene $\xrightarrow{(CH_3)_2NH}$ *N,N*-dimethyl-2-phenylethanamine $\xrightarrow{HNO_3}$ *bis*(methylene)[2-(3-nitrophenyl)ethyl]-λ^5-azane $\xrightarrow{HNO_2}$ 2-[2-(dimethylamino)ethyl]phenol

Physical Properties

Property	Value
Molecular Formula	= $C_{10}H_{15}NO$
Formula Weight	= 165.2322
Composition	= C(72.69%) H(9.15%) N(8.48%) O(9.68%)
Molar Refractivity	= 50.67 cm^3
Molar Volume	= 160.9 cm^3
Parachor	= 404.1 cm^3
Index of Refraction	= 1.542
Surface Tension	= 39.7 dyne/cm
Density	= 1.026g/cm^3
Polarizability	= 20.08 cm^3
Monoisotopic Mass	= 165.115364 Da
Nominal Mass	= 165 Da
Average Mass	= 165.2322 Da
Log P	= 1.40

Chemical Properties

On methylation by means of dimethyl sulphate, followed by oxidation with potassium permanganate in alkaline solution, hordenine is converted into anesic acid, when degraded by Hoffmann's method it yields trimethyl amine. The formula deduced by this method has been confirmed by the method of synthesis

Phenolic bases of this type but of higher molecular weight, in which the $(CH_3)_2$ group are further removed from the benzene ring than in hordenine, have also been synthesized. If the phenolic hydroxyl group is displaced fro the para hydroxyl group is displaced from the para to the ortho-position, the physiological action is much more weakened.

Uses

Hordenine sulphate raises the blood pressure and therefore used in medicine and it is also used therefore in the treatment of diarrhœa and dysentery.

EPHEDRINE

Ephedrine (EPH) is a sympathomimetic amine similar in structure to the synthetic derivatives amphetamine and methamphetamine. Ephedrine is commonly used as a stimulant, appetite suppressant, concentration aid, decongestant and to treat hypotension associated with regional anaesthesia. Chemically, it is an alkaloid derived from various plants in the genus *Ephedra* (family Ephedraceae). It is most usually marketed in the hydrochloride and sulfate forms.

OH
H
N
CH_3
CH_3

In traditional Chinese medicines, the herb *má huáng* (*Ephedra sinica*) contains ephedrine and pseudoephedrine as its principal active constituents. The same is true of other herbal products containing extracts from *Ephedra* species. Nagayoshi Nagai was the first one to isolate ephedrine from *Ephedra vulgaris* in 1885. The substance called soma mentioned in old Hindu books such as the Rig Veda, may have been ephedra extract. This, however, is disputed, as the identity of *soma*.

The production of ephedrine in China has become a multi-million dollar export industry. Companies producing for export extract 10 times the amount that is used in traditional Chinese medicine.

Physical Properties

Property	Value
Molecular Formula	$= C_{10}H_{15}NO$
Formula Weight	= 165.2322
Composition	= C(72.69%) H(9.15%) N(8.48%) O(9.68%)
Molar Refractivity	= 50.15 cm^3
Molar Volume	= 162.7 cm^3
Parachor	= 404.1 cm^3
Index of Refraction	= 1.528
Surface Tension	= 38.0 dyne/cm
Density	= 1.015 6 g/cm^3
Polarizability	= 19.88 cm^3
Monoisotopic Mass	= 165.115364 Da
Nominal Mass	= 165 Da
Average Mass	= 165.2322 Da
Log P	= 1.05

Chemical Properties

Ephedrine exhibits optical isomerism and has two chiral centres. By convention the enantiomers with opposite stereochemistry around the chiral centres are designated ephedrine, while pseudoephedrine has same stereochemistry around the chiral carbons. That is, (1*R*,2*R*)- and (1*S*,2*S*)-enantiomers are designated pseudoephedrine; while (1*R*,2*S*)- and (1*S*,2*R*)-enantiomers are designated ephedrine. The isomer which is marketed is (-)-(1*R*,2*S*)-ephedrine.

As with other phenylethylamines, it is also somewhat chemically similar to methamphetamine, although the amphetamines are more potent and have additional biological effects. Ephedrine may also be referred to as: (αR)-α-[(1S) 1(methylamino)ethyl] benzene methanol, α-[1-(methylamino) ethyl] benzyl alcohol, or L-erythro-2-(methylamino)-1 phenylpropan-1-ol. Ephedrine hydrochloride has a melting point of 187-188°C.

Uses

The use of Ephedrine for Human consumption, has been banned in the U.S. due to health risks, being that Ephedrine is a stimulant and at high does can cause Cardiac problems. The FDA banned Ephedrine from the side effects users were reporting, but the science seemed to be lost/not even considered and ephedrine was not tested in clinical trials. Ephedrine is not a Schedualed Drug. It is under Special Control under the Patriot Act **NOT**. The 1970 Controlled Substance Act, which states retailers in the

United States are required to collect the signatures of every person purchasing pseudoephedrine-containing products in log books and view a photo ID for every purchase of any product for which the entire box contains more than 60mg of pseudoephedrine. The "Combat Methamphetamine Epidemic Act of 2005" provides for criminal penalties for non-compliance and limits the maximum amount an individual can buy to 9 grams of pseudoephedrine per any 30-day period. This makes it unlawful for any person to knowingly or intentionally purchase at retail more than 9 grams during a 30-day period of which no more than 7.5 grams can be imported by private or commercial carrier or the Postal Service. The ban is also been trying to reach the Supreme Court but as of today they seem to be ignored.

ALKALOIDS OF THE PYRIDINE GROUP

α-CONIINE

Introduction and History

Coniine was the first of the alkaloids ever synthesized (by Albert Ladenburg in 1886).

Coniine is a poisonous alkaloid found in poison hemlock and the Yellow Pitcher Plant, and contributes to hemlock's fetid smell. It is a neurotoxin which disrupts the peripheral nervous system. It is toxic to all classes of livestock and humans; less than 0.2g (0.007oz) is fatal to humans, with death caused by respiratory paralysis. Socrates was put to death by way of this poison in 399 BC. Coniine has two streoisomers: (*S*)-(+)-coniine, which is the natural isomer present in hemlock and (*R*)-(-)-coniine.

Preparation

The synthesis of conüne was accomplished by Ladenberg with aid of the following three reactions (a) introduction of side groups into pyridine by heating pyridine alkaloids under pressure (b) condensation of α and γ methyl pyridines into piperidine by reduction with sodium and alcohol. The complete synthesis of coniine proceeds in the following stages:

Carbon disulphide, which can be prepared from its elements, is converted through the various intermediate compounds shown below into trimethyl bromide. The latter by way of trimethylene cyanide, yields pentamethylene diamine, from which piperidine is obtained by splitting of ammonia, Piperidine may be oxidized to pyridine, and this with methyl iodide gives the addition compounds pyridine methiodide, which at 300°C is transferred into hydriodide of α-picoline. On heating picoline itself with paraldehyde to a high temperature it gives α-allyl pyridine, which by reduction is converted into indicative coniine.

The reacemic base may be resolved means of δ-tartaric acid. On crystallizing a solution of *r*-coniine δ-tartrate, which is then removed and decomposed with alkali.

Synthetic coniine as thus obtained is identical in most respects with natural δ-coniine, from which it differs mainly in possessing a slightly higher rotation (by about 4%). The synthetic product was first believed to be an isomeric iso–coniine, which on heating alone or with alkali gave a product identical with the natural alkaloid. Recent work indicates that iso–coniine is merely an impure from of coniine.

carbon disulphide → carbon tetrachloride → tetrachloroethene → acetic acid → acetone

propan-2-ol → prop-1-ene → 1,2-dichloropropane → 1,2,3-trichloropropane → glycerol

1,3-dibromopropane → 3-chloropentanedinitrile → pentane-1,5-diamine → piperidine → pyridine → 1-iodo-1-methyl-1 l^5-pyridine

2-methylpyridine → 2-prop-1-en-1-ylpyridine → (2R)-2-propylpiperidine dl → Resolution → iso coniine

d - coniine

Physical Properties

Property	Value
Molecular Formula	$= C_8H_{17}N$
Formula Weight	= 127.22728
Composition	= C(75.52%) H(13.47%) N(11.01%)
Molar Refractivity	= 40.36 cm^3
Molar Volume	= 156.5 cm^3
Parachor	= 356.9 cm^3
Index of Refraction	= 1.429
Surface Tension	= 27.0 dyne/cm

Density	= 0.812 g/cm^3
Polarizability	= 16.00 10^{-24}cm^3
Monoisotopic Mass	= 127.1361 Da
Nominal Mass	= 127 Da
Average Mass	= 127.2273 Da
Log P	= 2.49

Chemical Properties

A. W Hoffmann's subjected coniine to certain reactions which he has previously carried out with piperidine, and found that two compounds behaved in a similar manner.

1. **Exhaustive Methylation of** coniine gave a product having the composition of a dimethyl-coniine, $C_8H_{15}(CH_3)_2N$ and also a hydrocarbon conylene) of the formula C_8H_{14}

$$C_8H_{17}N \rightarrow C_8H_{16}(CH_3)NCH_3I \rightarrow C_8H_{16}(CH_3)NCH_3OH \xrightarrow{-H_2O} C_8H_{15}(CH_3)_2N \rightarrow C_8H_{15}(CH_3)_3N \rightarrow$$
$$\rightarrow C_8H_{15}(CH_3)_3N.OH \rightarrow C_8H_{11} + N(CH_3)_3 + H_2O + \text{Coniine}$$

Hoffmann observed that differed from piperylene by the same atomic complex C_3H_6 as coniine from piperidine, and hence of piperidine. This view was confirmed by Hoffmann's distillation of the alkaloid with the zinc dust.

2. **The distillation with zinc dust** under taken in the expectation of obtaining from coniine a compound richer in hydrogen. In actual practice it was found that hydrogen was removed and the compounds $C_8H_{17}N$. The new base conyrine, was easily recognized as a derivative of pyridine and being six atoms poorer in hydrogen than coniine appeared to stand to the latter in the same relationship as pyridine to piperidine. Any doubt as to the nature of conyrine was resolved by its conversion into picolinic acid or 2-pyridine-carboxylic acid on oxidation.

From this it followed that conyrine must be either 2-propyl or 2-isopropyl pyridine, and coniine therefore be 2-propyl or 2-isopropyl-piperidine. The choice between these two alternatives was decided in favour of the normal propyl structure by Hoffmann's discovery that coniine, on reduction with hydrochloric acid, gave ammonium and normal octane. Has an isopropyl group been present this could not have occurred without intermolecular rearrangement.

The constitution of coniine as 2 propyl-piuperidine was finally confirmed by synthesis

3. **The oxidation of coniine** with hydrogen peroxide led to the formation of 8-propyl-δ-amminovaleraldehyde (δ-amino–*n*–octic aldehyde).

PIPERINE

Introduction

Piperine was first discovered by Hans Christian Ørsted in 1819.

Piperine is the alkaloid responsible for the pungency of black pepper along with chavicine (an isomer of piperine). It has also been used in some forms of traditional medicine and as an insecticide.

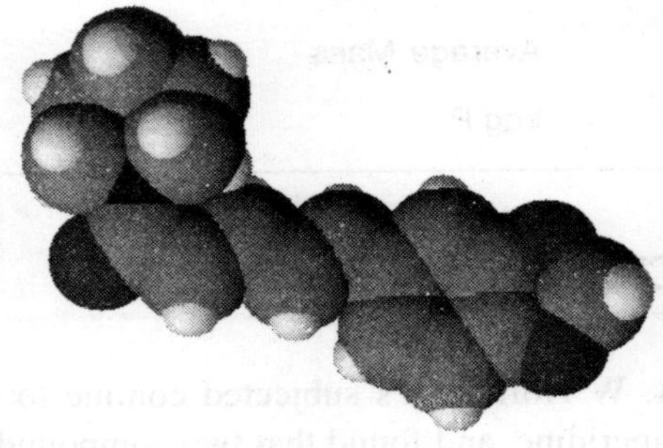

The pungency caused by capsaicin and **piperine** is caused by activation of the heat and acidity sensing TRPV ion channel TRPV1 on nociceptors (pain sensing nerve cells).

Piperine has also been found to inhibit human CYP3A4 and P-glycoprotein, enzymes important for the metabolism and transport of xenobiotics and metabolites. In animal studies, piperine also inhibited other enzymes important in drug metabolism. By inhibiting drug metabolism, piperine may increase the bioavailability of various compounds. Notably, piperine may enhance bioavailability of curcumin by 2000% in humans.

Due to its effects on drug metabolism, piperine should be taken cautiously (if at all) by individuals taking other medications.

Physical Properties

Property	Value
Molecular Formula	= $C_{17}H_{19}NO_3$
Formula Weight	= 285.33766
Composition	= C(71.56%) H(6.71%) N(4.91%) O(16.82%)
Molar Refractivity	= 82.14 ± 0.3 cm^3
Melting point	=130 °C
Molar Volume	= 235.4 cm^3
Parachor	= 629.2 cm^3
Index of Refraction	= 1.614
Surface Tension	= 51.0 dyne/cm
Density	= 1.211 g/cm^3
Polarizability	= 32.56 cm^3
Monoisotopic Mass	= 285.136493 Da
Nominal Mass	= 285 Da
Average Mass	= 285.3377 Da
Log P	= 2.66

Chemical Properties

When boiled with alcoholic potassium hydroxide breaks down into piperidine and piperic acid.

piperine + H_2O → piperidine + piperic acid

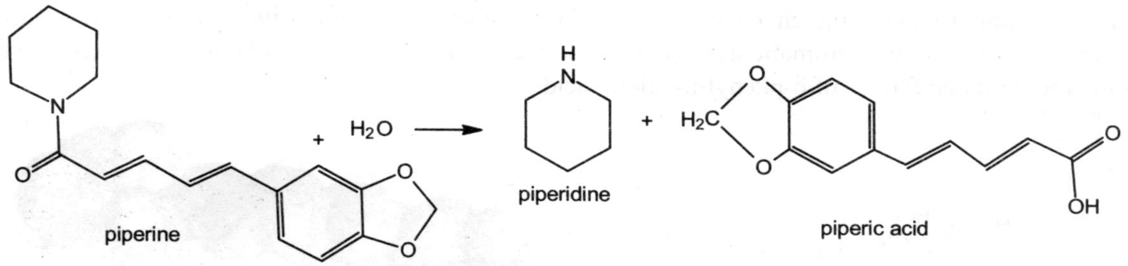

Hence it was concluded that piperine is a compound of amide type built up from piperidine and piperic acid. This view was confirmed by the partial synthesis of piperine on heating piperidine in benzene solution with the chloride of piperic acid. This view was confirmed by the partial syntheisis of piperine on heating piperidine in benzene solution. With the chloride of piperic acid.

piperidine + piperic acid chloride → piperine

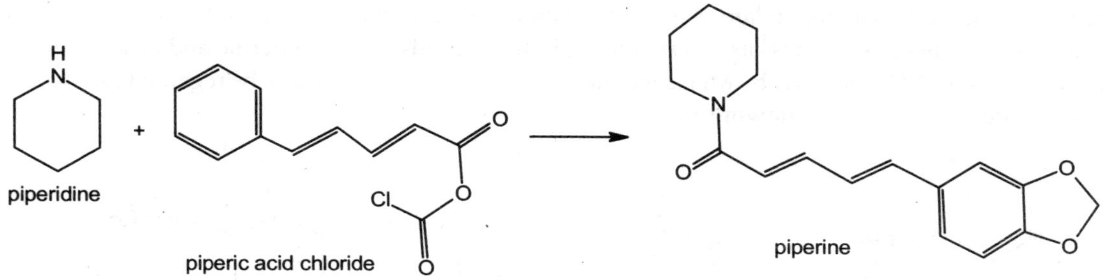

The constitution and the syntheisis of piperidine have been described previously in the chapter of heterocyclic compounds and the structure of piperic acid was solved by Fittig and confirmed by the following synthesis of Ladenberg and Scholtz. Piperonal was condensed with acetaldehyde in the presence of aqueous alkali to give the unsaturated aldehyde, piperonyl acrolein; the latter was when converted into piperic acid by the use of Perkin's reaction.

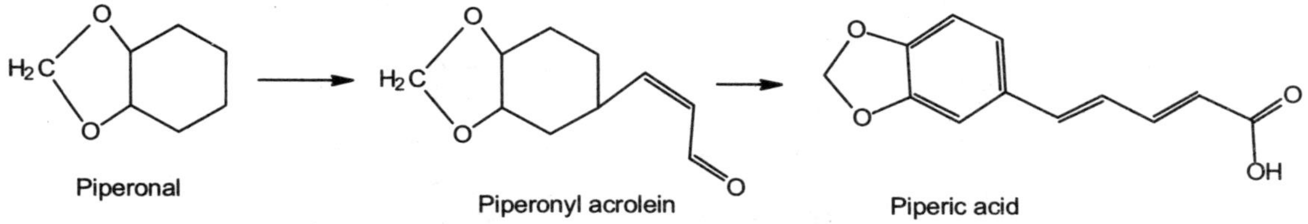

Piperonal → Piperonyl acrolein → Piperic acid

Consequently the above preparation of piperine from its hydrolysis products piperidine and piperic acid completes the synthesis of this alkaloid.

Owing to the rising price of pepper, experiments have been directed towards an artificial product of similar taste. An actual synthesis of piperine is out of the question owing to the cost of starting material by information as to the relationship between constitution and pepper-like taste has been gained by the work of H. Staudinger. It appears that the molecule of piperine may undergo considerable changes without loosing the characteristic taste. An essential condition is the acid amide linking of piperidine with a fatty – aromatic acid radical, and the most pronounced resemblence to pepper was observed with derivatives of δ-phenyl-n-valeric acid.

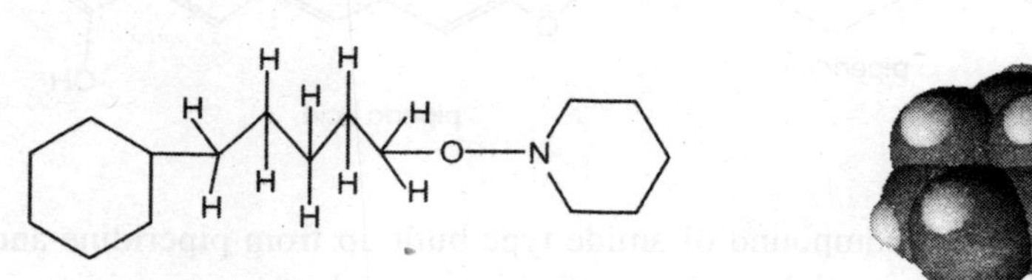

Alkaloids of Pomegranate Black

The bark of pomegranate tree (*Punica granatum*) contains several alkaloids, to the presence of which is due its long-known usefulness as a vermicide. These alkaloids viz. Pelletierine and isopelleteirine of the formula $C_8H_{15}NO$, and latter by Meisenheimer. Their constitution, can now be regarded as established in the accordance with the following formula.

H N, CH_2—CH_2, CH=O

3-piperidin-2-ylpropanal
Pelletierine

H N, CH_2—CH_2, C=O, CH_3

4-piperidin-2-ylbutan-2-one
isopelletierine

CH_3 N, CH_2—CH_2, C=O, CH_3

4-(1-methylpiperidin-2-yl)butan-2-one
Methyl isopelletierine

Pseudo-pelletierene recently been synthesized by Menzies and Robinson in a simple manner from glutaric acid aldehyde as follows (compare) synthesis of tropinone.

CH_3, CH_2, CH_2, O

Glutardialdehyde

\+

H, H N—CH_3

methanamine

\+

H_2C, O, Ca^+, C=O, CH_2, Ca^+ O, O

Calcium acetone dicarboxalate

Pseudo pelletierine

Alakloids of the Pyrolidine Group and Derivatives of Tropane

In this group are included, nicotine, atropine hyoscyamine, cocaine, topacocaine and others. Since the five membered pyrolidine ring is more easily formed than the corresponding six-membered ring the production of alkaloids of the pyrrolidine type in plants is not surprising. In all probably a number of the other alkaloids the constitution of which is still unknown will eventually be found to fall within this class.

NICOTINE

Introduction and History

Nicotine is named after the tobacco plant *Nicotiana tabacum*, which in turn is named after Jean Nicot, French ambassador in Portugal, who sent tobacco and seeds from Brazil to Paris in 1550 and promoted their medicinal use. Nicotine was first isolated from the tobacco plant in 1828 by German chemists Posselt & Reimann. Its chemical empirical formula was described by Melsens in 1843, and it was first synthesized by A. Pictet and Crepieux in 1893. Nicotine was originally banned for use in the United States but was brought in to the country as a horse tranquilizer then illegally transferred to tobacco farms for use in their product. Nicotine is an alkaloid found in the nightshade family of plants (*Solanaceae*), predominantly in tobacco, and in lower quantities in tomato, potato, eggplant (aubergine), and green pepper. Nicotine alkaloids are also found in the leaves of the coca plant. Nicotine has been found to constitute approximately 0.6-3% of dry weight of tobacco, with biosynthesis taking place in the roots, and accumulating in the leaves. It functions as an antiherbivore chemical, being a potent neurotoxin with particular specificity to insects; therefore nicotine was widely used as an insecticide in the past, and currently nicotine derivatives such as imidacloprid continue to be widely used.

In low concentrations (an average cigarette yields about one mg of absorbed nicotine), the substance acts as a stimulant in mammals and is one of the main factors responsible for the dependence-forming properties of tobacco smoking. According to the American Heart Association, "Nicotine addiction has historically been one of the hardest addictions to break." The pharmacological and behavioral characteristics that determine tobacco addiction are similar to those that determine addiction to drugs such as heroin and cocaine.

Synthesis of Nicotine

Starting from β-aminopyride, the synthesis of nicotine involved the following steps

β-Amino pyridine $\xrightarrow{\text{distillation with music acid}}$

I-β-Pyridyl pyrrole $\xrightarrow{\text{Strongly heated}}$

2β-Pyridyl pyrrole $\xrightarrow{\text{potassium salt with methyl iodide}}$

I-Methyl-2-β-pyridyl-pyrrolemethiode $\xrightarrow{\text{distillation with lime}}$

I-Methyl-2-β-pyridyl-pyrrole (nicotyrine) $\xrightarrow{\text{action of I}}$

Iodo nicotyrine (1-methyl-2-β-pyridyl-4-iodopyrrole $\xrightarrow{\text{reduction}}$

Dihydro-nicotyrine (1-methyl-2-β-pyridyl-pyrroline $\xrightarrow[\text{with Sn and HCl}]{\text{reduction of the perbromide}}$

Tetrahydro-nicotyrine (dl-nicotine) $\xrightarrow[\text{tartaric acid}]{\text{resolution with}}$ *l*-nicotine

A further synthesis was subsequently effected by E.Späth and H.Bretschnieder in the following steps:[1]

ethyl nicotinate + 1-methylpyrrolidin-2-one → 3-benzoyl-1-methylpyrrolidin-2-one $\xrightarrow{\text{heat with HCl}}$ N-methyl-4-phenylbutan-1-amine

$\xrightarrow{\text{reduction with zinc dust}}$ N-methyl-4-phenylbutan-1-amine $\xrightarrow{\text{HI}}$ 4-iodo-N-methyl-4-phenylbutan-1-amine $\xrightarrow{\text{NaOH}}$ 1-methyl-2-phenylpyrrolidine

[1]The first condensation is this synthesis is brought about in the presence of sodium ethoxide, giving a product which, in being heated with fuming hydrochloric acid, is hydrolysed with loss of carbon dioxide.

l – NICOTINE

Introduction

The naturally occurring alkaloid is lævorotatory and as indicated above, can be also obtained by resolving the synthetic *dl*-nicotine with the aid of tartaric acid. According to the kind of tobacco the nicotine content varies from 0.6% to 0.8%. In general the finer kinds of tobacco contains lesser amount of nicotine. The alkaloid is conveniently obtained from extract of tobacco, which is prepared industrially by extracting a raw tobacco of high nicotine content with cold water and concentrating the solution. The extract is used for the impregnating of chewing tobacco, and contains about 8% to 10% of nicotine. It is first diluted with water and freed form hydrocarbons by the additions of acids and extraction with ether. The solution is then made alkaline, and the free nicotine repeatedly extracted with ether.

Freshly prepared *l*-nicotine is a colourless oil, which dissolves readily in water, has a burning taste, and is very poisonous, when pure, it has and unpleasent, stupefying[1] odour unlike that of tobacco it can only by distilled without decomposition in a current of hydrogen; in air it rapidly turns brown and refines, Nicotine boils at 246.2°C.

Physcial Properties

Molecular Formula	= $C_{10}H_{14}N_2$
Formula Weight	= 162.23156
Composition	= C(74.03%) H(8.70%) N(17.27%)
Melting point	= -79 °C
Boiling point	= 247 °C
Molar Refractivity	= 49.25 cm^3
Molar Volume	= 157.1 cm^3
Parachor	= 394.2 cm^3
Index of Refraction	= 1.539
Surface Tension	= 39.6 dyne/cm
Density	= 1.032 g/cm^3
Polarizability	= 19.52 cm^3
Monoisotopic Mass	= 162.115698 Da
Nominal Mass	= 162 Da
Average Mass	= 162.2316 Da
Log P	= 0.72

[1]Astonishing.

Chemical Properties

It forms diacid salts which do not crystallize well; these dissolve readily in water and rotate the plane of polarization to the right.

Nicotine yields two methiodides. One of these isomerides is obtained as a syrupy mass on bringing together equimolecular amounts of nicotine and methyl iodide. The second results when nicotine is first treated with a molecular equivalent of hydroiodic acid and then with less basic nitrogen atom of the pyridine ring, by converting oxidizing the latter with potassium permanganate.

Pictet obtained the alkaloid trigonelline, which is present in the fenugreek* of Stropahnthus hispadus and other plants.

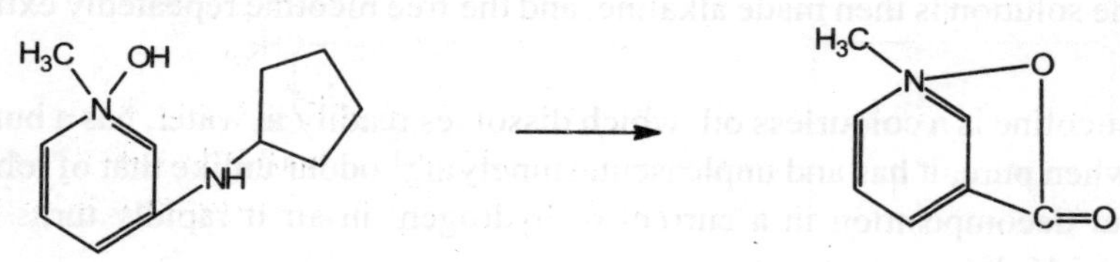

3-(cyclopentylamino)-1-methyl-1 λ^5-pyridin-1-ol 1-methyl-7-oxa-1 λ^5-azabicyclo[3.2.1]octa-1(8),2,4-trien-6-one

trigonelline methyl betaine

Conversion of *l* - Nicotine Into *dl* - Nicotine

On heating an aqueous solution of monohydrochloride or sulphate of nicotine at 180 to 250°C in a sealed tube, the rotation steadily diminishes and finally becomes zero.

d-l Nicotine may be isolated form heated solutions of its salts in the usual manner. In properties such as boiling point, specific gravity, refractive index, smell, solubility and salt formation it is identical with the natural-*l*-rotatory alkaloid.

d - NICOTINE

This was isolated in the crude state during the preparation of *l*-nicotine from the inactive synthetic compound, and purified by use of *l*-tartaric acid. Its specific rotation $[\alpha]\frac{20}{D}$ was found to be 163.17 and in boiling point and other physical properties it was identical with the *l*-isomeride.

---

*Fenugreek (*Trigonella foenum-graecum*) or menthya (Kannada) or Venthayam (Tamil) or menthulu (Telugu) belongs to the family Fabaceae. Fenugreek is used both as a herb (the leaves) and as a spice (the seed). It is cultivated worldwide as a semi-arid crop. The name fenugreek or *foenum-graecum* is from Latin for "Greek hay". Zohary and Hopf note that it is not yet certain which wild strain of the genus *Trigonella* gave rise to the domesticated fenugreek but believe itwas brought into cultivation in the Neart East. Charred fenugreek seeds have been recovered from Tell Halal, Iraq, (radiocarbon dating to 4000 BC) and Bronze Age levels of Lachish, as well as desiccated seeds from the tomb of Tutankhamen. Cato the Elder lists fenugreek with clover and vetch as crops grown to feed cattle.

δ-Nicotine is less poisonous than *l*-nicotine. In this respect, the different action of the two antipodes towards the animal organism may be compaired to the different behaviour of optical antipodes in general towards any other optically active compound and towards organized as distinct from unorganized, formats.

COMPOUNDS OF TROPANE SERIES

Nomenclature

The various alkaloids of this group contain a peculiar combination of a reduced pyrrole and a reduced pyridine ring, the periphery of the cyclic system forming a seven membered ring of carbon.

(I) (II)

Derivation of tropane are generally described by the use of numbering given in formula the compounds being preferred in the customary manner to tropane (II) as parent substance.

Willstätter system of synthesis of tropane and tropane derivatives are based upon the alkyalting action of a halogenated group on a basic group of same molecule.

Just as an alkyl halide results with a primary amine to yield the salt of a secondary amine, or with a halogenated base an intramolecular reaction may occur between the halogenated portion of the molecule and the basic group. In such a case the halogen atom and the alkyl residue to which it was originally united become attached by separate valency bonds to the nitrogen atom, leading to the production of a cyclic base in which nitrogen forms part of the ring. Thus a halogenated primarily base yields the salt of an imine and a tertiary compound yields a quaternary ammonium halide. A reaction of this type is described by Willstätter as intermolecular alkylation (compare the synthesis of 2-methyl-1-dimethyl-pyrrolidium chloride from phenyl-dimethyl-amine.

Intermolecular alkylation may also lead to the formation of dicyclic bases if the addition products of certain unsaturated monocylic amines are as starting material.

Derivatives of tropane have been synthesized in this manner by Willstätter, starting from a base containing a cyclic system of seven carbon atoms, and having a halogen atom in one of the two δ - position, (i.e. positions 4 and 5) to the N-group.

Tropane, Hydrotropidine

The parent substance of the tropane series, was first obtained by Ladenburg by the action of zinc dust and hydrochloric acid on tropine iodide

$$C_6H_{14}IN, HI + 2H \rightarrow C_8H_{15}N, 2HI$$

According to Willstätter tropane is best prepared from the hydrogen halide addition products of tropidine by reduction with zinc dust in the cold. It is also formed from tropinone by treatment with zinc dust and hydroiodic acid. Willstätter has synthesized tropane by two methods. One of these starts from the addition compounds obtained by union of Δ^4 dimethyl-amino-cyclopentane with hydrochloric acid.

tropane

When this is gently heated it is largely converted by intermolecular ammonium salt formation into tropane methochloride, from which, on dry distillation tropane is obtained. It is a liquid of boiling point 167°C; which is sparingly soluble in cold and still less soluble in hot water.

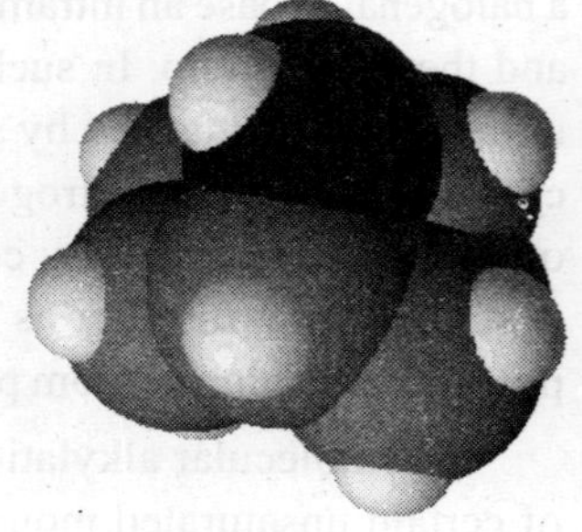

On exhaustive methylation tropane yields hydrotropilidine or cyclo-heptadiene $C_{17}H_{10}$, the final stage of the degradation being as follows:

$$C_{17}H_{11}N(CH_3)_3OH \rightarrow C_{17}H_{10} + N(CH_3)_3 + H_2O$$

Nortropane or hydrotropidine, $C_7H_{13}N$ is formed by distilling tropane in a stream of hydrochloric acid, when the N-methyl group is removed as methyl chloride (Ladenburg).

8-methyl-8-azabicyclo[3.2.1]octane hydrochloride
tropane hydrochloride

8-azabicyclo[3.2.1]octane
Nortropane

Physical Properties

Nortropane is a transparent crystalline substance boiling about 161°C and melting at 60°C

Property	Value
Molecular Formula	$= C_7H_{13}N$
Formula Weight	= 111.18482
Composition	= C(75.62%) H(11.79%) N(12.60%)
Molar Refractivity	= 33.68 cm^3
Molar Volume	= 118.9 cm^3
Parachor	= 281.8 cm^3
Index of Refraction	= 1.478
Surface Tension	= 31.4 dyne/cm
Density	= 0.934 g/cm^3
Polarizability	= 13.35 $10^{-24} cm^3$
Monoisotopic Mass	= 111.104799 Da
Nominal Mass	= 111 Da
Average Mass	= 111.1848 Da

Chemical Properties

When distilled with zinc dust it yields 2-ethyl pyridine a reaction which first led to the discovery that tropine contains a pyridine nucleus.

TROPINE

Introduction

Tropine, the basic cleavage* product of most of the Solanaceæ alkaloids (e.g. atropine) is one of the most important derivatives of tropane. It has been more completely investigated than any of the other derivatives, which results with gave the first insight into the structure of the tropane ring.

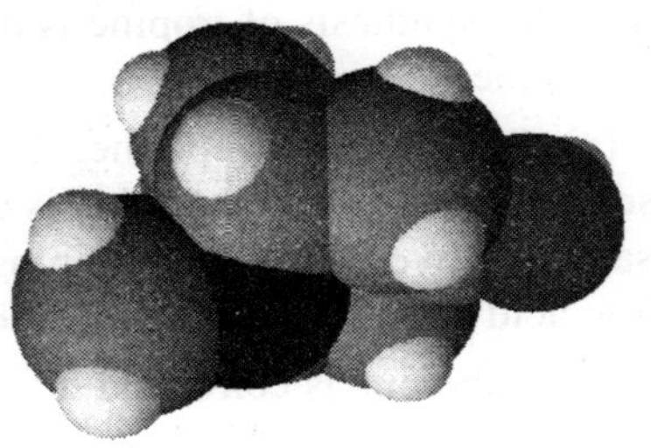

*A crystalline compound comparatively lesser complex in nature than that of its parent compound, got chemical separated.

Formation and Properties of Tropine

Tropine was first obtained by the hydrolysis of atropine with barium hydroxide (Kraut 186) latter isolated in a manner from hyoscyamine (Ladenburg) and belladonine (Merling, Willstätter) prepared tropine by the reduction of tropinone and finally affected its synthesis.

The base, which is optically inactive, crystallizes from absolute ether in large plates with melting point 63°C and b.p 229°C. It dissolves readily in water and alcohol, giving solutions with a strong alkaline reaction.

Ψ - Tropine, described later, is a geometrical isomeride of tropine

Cosntitution of Tropine

Following are the main points from which the constitution of tropine is based upon the transformation of the latter into tropidine by simple removal of water. Since tropine is a tertiary base and thus contains no hydrogen, it must be the hydrogen of its hydroxyl group which is replaced by an acidic radical in the alkaloid atropine.

That atropine contains a pyridine nucleus follows from the conversion of tropidine into di bromo-pyridine and α-ethyl-pyridine.

The presence of a seven membered carbon ring in tropane is shown by the conversion of tropane is shown by the conversion of tropidine into tropilidene or cycloheptriene on exhaustive methylation and also by the degradation of tropionic acid to normal pimelic acid. Willstätter has conclusively proved the existence of a pyrolidine nucleus in tropine by an examination of the degradation products fromed on oxidation. Tropionic acid was identified as 1-methyl–pyrolidine-2-carboxylic-5-acetic acid and by energetic oxidation was converted into N-methyl-succinimide. In this manner the pyrrolidine nucleus was isolated from the tropine in a simple well known form.

In establishing the above formula for tropidine, a factor of importance was the observation that tropinone, the primary oxidation product of tropine, readily yields a dibenzal compound and a di-isonitrous derivative and must therefore contain the group $CH_2 - CO - CH_2$.

Synthesis of Tropine

Willstätter synthesis of tropine is divided into two parts the synthesis of tropidine and its conversion into tropine.

(a) Synthesis of tropidine, this synthesis has been affected in two ways, only one of which is described here. In nut shell this is the reversal of the stages by which tropine may be degraded to an unsaturated hydrocarbon containing of seven carbon atoms. The starting is suberone, obtained from subric acid and the synthesis proceeds in the following steps:

I. Suberone is converted into cycloheptane and then into cylcoheptadiene and cylcoheptatriene.

II. Cylcoheptatrine is converted into dimethylamino-cylcoheptatriene

III. The hydrogen halide addition product of this monocylcic tropine base is transformed into a dicyclic tropane methyl-ammonium salt, which on distillation yields tropedine.

Synthesis of Cylcoheptatriene

Suberic acid, which can be obtained from glutaric acid by the electrolytic method of Crum Brown and Walker, is converted into the calcium salt and distilled. The suberone or cycloheptanone, prepared in this way, it was first converted into the hydro carbon cycloheptane, containing one double bond.

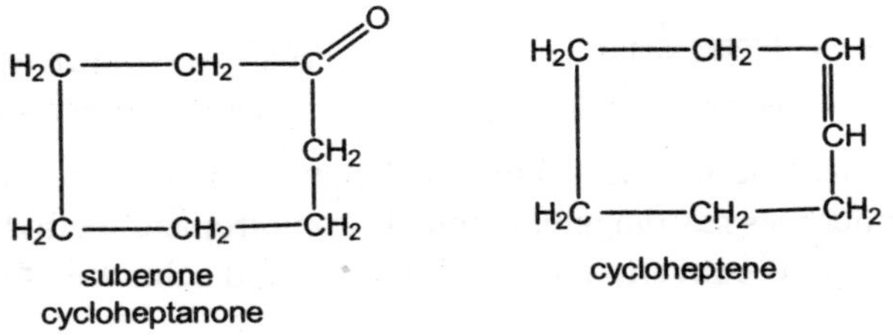

suberone
cycloheptanone

cycloheptene

This can be accomplished either by treating suberyl iodide with alcoholic potash (Markowinkoff, or by the exhaustive methylation of suberylamine (amino cycloheptane) obtained by the reduction of suberone oxime (Willstätter).

A second double bond is introduced into the molecule by allowing cycloheptane and unsaturated Δ^2 dimethylamino – cylcoheptene is formed, according to the equation:

1,2-dibromocycloheptane + 2 HN(CH_3)$_2$ (*N*-methylmethanamine) ⟶ bromo(cyclohept-2-en-1-yl)dimethylammonium + HN(CH_3)$_2$ (*N*-methylmethanamine) + HBr

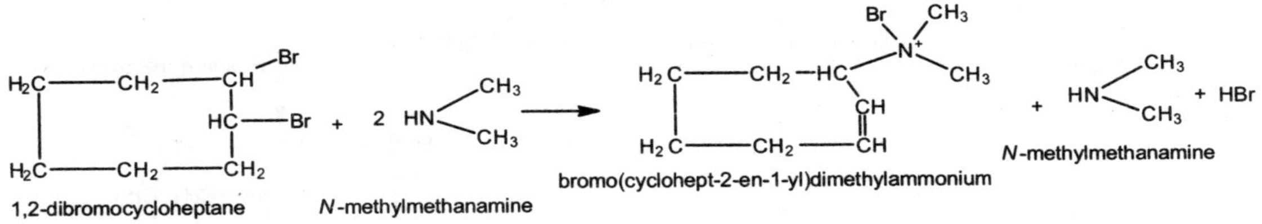

This dibromide of cycloheptadiene can be converted into cylcoheptatriene by various methods, when head with quinoline, for example, hydrogen bromide is removed and a quantitative yield of cylcoheptatriene is obtained.

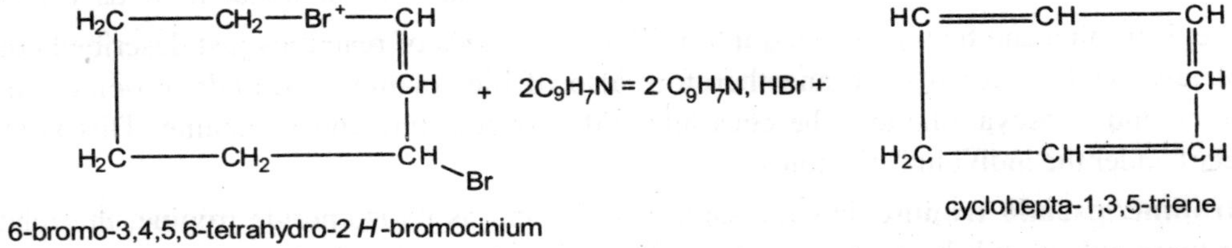

6-bromo-3,4,5,6-tetrahydro-2 *H*-bromocinium + $2C_9H_7N = 2\ C_9H_7N,\ HBr$ + cyclohepta-1,3,5-triene

The synthetic cycloheptatriene produced this way from suberone is identical in all respects with the tropilidene prepared from tropine.

2. Conversion of Cycloheptatriene hydro bromide, which is formed by treating the hydrocarbon in the cold with one molecular property of the hydrogen bromide, reacts readily at ordinary temperature with dimethyl amine in benzene solution, with the production of dimethyl amino-cylcoheptadiene, the doubly unsaturated base taking upto two atoms of hydrogen according to the equation.

N,N-dimethylcyclohepta-2,4-dien-1-amine → N,N-dimethylcyclohept-4-en-1-amine

Dimethylamino-cylcoheptene in acid solution adds on bromine to form a dibromide, which, on being warmed, rapidly undergoes rearrangement into 4-bromotropane-methyl readily decomposes into hydrobromic. When the latter is converted into the chloride and submitted to dry distillation and then yields tropidine.

→ 4-bromo- N,N-dimethylcyclohept-4-en-1-amine → tropidine methyl bromide → tropidine

Tropidine may be converted into Ψ-tropine by way of its hydrogen bromide addition compound, bromotropane, when this is heated to 200°C with sulphuric acid in a sealed tube it yields Ψ-tropic acid.

Since Ψ-tropine can be transformed into tropine, the series of reactions just described constitutes a completely synthesis of tropine and therefore also of the solanine alkaloids atropine, atropanine belladonine and hyoscyamine and the coca alkaloids tropacocaine and *r*–cocaine. This is treated in more detail under the individual alkaloids.

Ψ-tropine, pseudo tropine, has the same constitution as its isomeride tropine, the relationship between these compounds being of the cis – between borneol and isoborneol.

Further examples of these same kinds will be met in tropine series. The relationship may be illustrated by the following space formula.

Ψ-tropine boils at 240°C to 241°C and crystalline needles of melting point 108°C. It is optically inactive, and readily dissolves in alcohol and water to give strongly alkaline solution. The base can be identified and separated from tropine by means of its *picrate* $C_8H_{15}NO$, $C_6H_2(NO_2)_3OH$.

Ψ-tropine was discovered at much latter date than tropine and is less prepared. It can be obtained from tropine in two ways, viz, directly, by heating it with a solution of sodium ethoxide and alcohol.

TROPINONE

Introduction

Tropinone is an alkaloid, famously synthesized in 1917 by Robert Robinson as a synthetic precursor to atropine, a scarce commodity during World War I. Tropinone and the alkaloids cocaine and atropine all share the same tropane core structure.

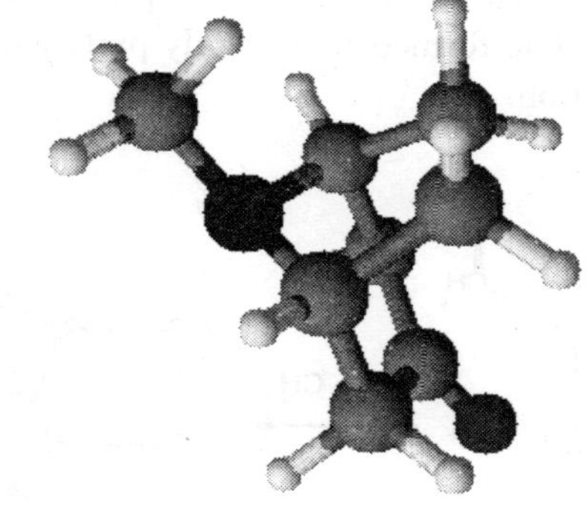

Preparation

The following synthesis of tropinone was carried out by Willstätter N-Methyl pyrallole-2:5-diacetic ester is reduced with hydrogen in presence of oxygenated platinum black to give N-methyl pyrrolidine-2 : 5 diacetic ester. The latter with sodium ethoxide undergoes intermolecular acetoacetic ester condensation to form tropinone carboxylic acid ester (II) which on warming with mineral acids gives tropinone.

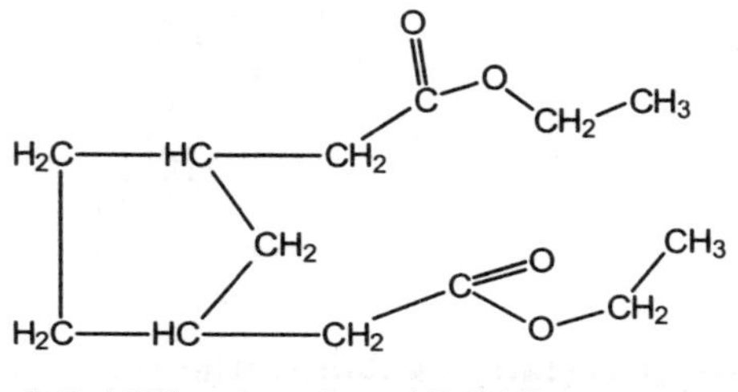

diethyl 2,2'-cyclopentane-1,3-diyldiacetate

(I)

4-ethyl-9-methyl-9-azabicyclo[4.2.1]nonan-3-one

(II)

Robinson's Synthesis of Tropinone

A remarkably simple synthesis of tropinone has been derived by Robinson Succinaldoxime and nitrous fumes – was allowed to interact in aqueous solution with acetone and methyl amine. After the lapse of half an hour at the ordinary temperature tropinone was found to be present.

succinaldehyde + $H_2N{-}CH_3$ + acetone → tropinone + $2H_2O$

In an experiment in which the calcium salt of acetone dicarboxylic acid was employed in place of acetone, a yield to tropinone amounting to about 42% was obtained. The tropinone dicarboxylate formed first readily parts with the two molecules of carbon dioxide on being heated in acid solution.

succinaldehyde + $NH_2{-}CH_3$ → 1-methylpyrrolidine-2,5-diol + calcium acetone dicarboxylate → tropinone

Robinson's theory of the Phytochemical Synthesis of certain alkaloids: The synthesis of tropinone described above proceeds with such case that Robinson has been suggested that similar in the plant. All knew carbon to carbon links are assumed to result from (1) an aldol condensation, or (2) the closely related condensation between a compound containing the group: CH.CO and a carbinol amine having the group C(OH).N, products of the latter type being readily obtained by union of an aldehyde or ketone with ammonia or an ammine.

The essential starting points in plant synthesis may be ammonia and formaldehyde, amino acids such as ornithine or lysine and certain degradation products of carbohydrates. For examples formaldehyde is known to exert a combined methylating as oxidizing action on ornithine leading to the loss of ammonia, and carbon dioxide and the formation of which may undergo cyclisation to give a carbinol and carbinol amine, derived from pyrrolidine.

$$NH_2-CH_2.CH_2.CH_2.CHNH_2.CO_2H + CH_2O \rightarrow NH-CH_3.CH_2.CH_2.CH_2.CHO + NH_2 + CO_2 \rightarrow$$

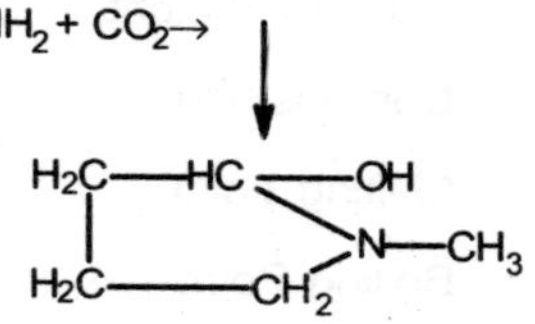

Further interaction with formaldehyde is supposed to attack both ends of the amino – acid molecule, yielding succindialdehyde and methyl amine, which combine to form the dihydroxy base II.

$$NH_2.CH_2\ CH_2\ CH_2\ CH.NH_2CO_2H + 2CH_2O \rightarrow OCH.\ CH_2.\ CH_2.CHO + 2\ CH_3NH_2 + CO_2$$

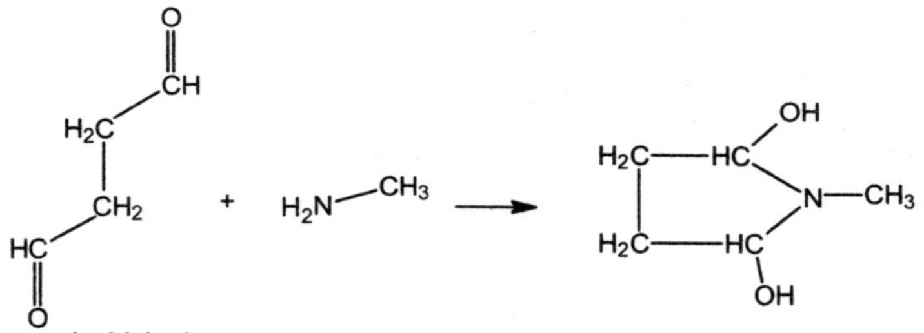

succinaldehyde

Among carbohydrate disruption products citric acid is suggested as providing acetone residues in the form of acetone dicarboxylic acid. Alternately, the latter compound is known to be proved *in vitro* form other sources e.g. by the spontaneous decomposition of calcium tri-saccarate.

Condensation is then assumed to proceed along the following lines, according to which on molecule of acetone-dicarboxylic acid reacts with either one or two molecules of base I. Subsequent elimination of carbon dioxide leads in the former case to hygrine (III) and in the latter to ***cuskygrine***

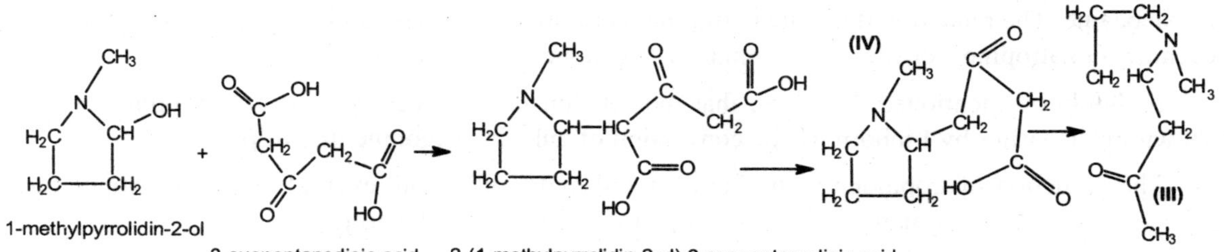

1-methylpyrrolidin-2-ol + 3-oxopentanedioic acid → 2-(1-methylpyrrolidin-2-yl)-3-oxopentanedioic acid → (IV) → (III)

Physcial Properties

Molecular Formula	= $C_8H_{13}NO$
Formula Weight	= 139.19492
Composition	= C(69.03%) H(9.41%) N(10.06%) O(11.49%)
Melting point	= 42.5 °C
Boiling Point	= 224°C
Molar Refractivity	= 38.73 cm^3
Molar Volume	= 130.5 cm^3
Parachor	= 318.7 cm^3
Index of Refraction	= 1.505
Surface Tension	= 35.5 dyne/cm
Density	= 1.066 g/cm^3
Polarizability	= 15.35 $10^{-24}cm^3$
Monoisotopic Mass	= 139.099714 Da
Nominal Mass	= 139 Da
Average Mass	= 139.1949 Da

Chemical Properties

Behaviour of Tropinone on Reduction

The best results were obtained by reduction of tropinone in the cold with zinc dust and hydroiodic acid. In this way a good yield of tropine, together with a smaller amount of Ψ-tropine, it is through tropinone into tropine, a change which cannot ne effected in any other manner, and which has been of in connection with the synthesis of tropine, atropine and other compounds.

Tropinone has also been obtained by the oxidation of ecgonine a hydrolysis product of cocaine and tropacocaine. The reduction if ketone to tropine therefore makes is possible to convert tropacocaine or cocaine into atropine Cocaine → ecgonine → tropinone → tropine → atropine.

It should be mentioned, however, that the relationship between cocaine and atropine has been shown much earlier by Einhorn by the conversion of anhydro-ecgonine into tropine.

The reduction of tropinone with zinc dust and hydroiodic acid, even at very low temperatures, proceeds beyond the formation of the alcohol base and finally yields tropane.

When reduced with sodium in moist ethereal solution tropinone is converted into Ψ-tropine, the same being obtained by use of sodium amalgam acidic solution of a dilute condition.

ECGONINE

Introduction

Ecgonine ($C_9H_{15}NO_3$), is an organic chemical found naturally in coca leaves. It is has a close structural relation to cocaine: it is both a metabolite and a precursor, and as such, it is a controlled substance, as are all known substances which can be used as precursors to ecgonine itself.

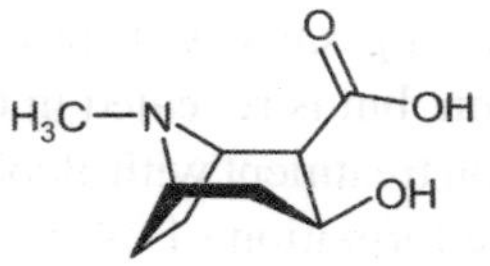

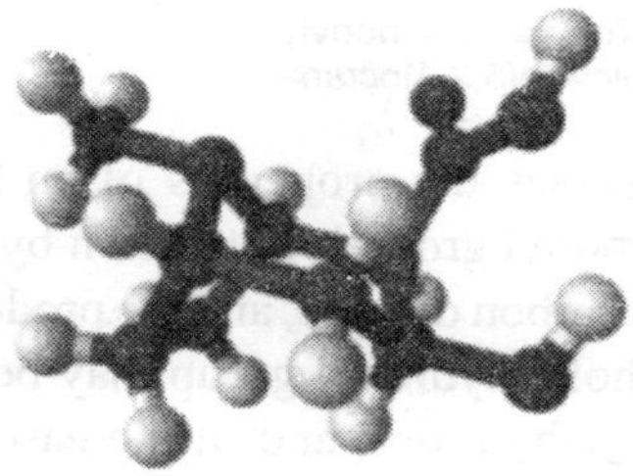

Structurally, ecgonine is a cycloheptane derivative with a nitrogen bridge. It is obtained by hydrolysis of cocaine with acids or alkalis, and crystallizes with one molecule of water, the crystals melting at 198–199°C. It is levorotary, and on warming with alkalis gives iso-ecgonine, which is dextrorotary.

It is a tertiary base, and has the properties of an acid and an alcohol. It is the carboxylic acid corresponding to tropine, for it yields the same products on oxidation, and by treatment with phosphorus pentachloride is converted into anhydroecgonine, $C_9H_{13}NO_2$, which, when heated to 280°C with hydrochloric acid, eliminates carbon dioxide and yields tropidine, $C_8H_{13}N$.

Constitution of Ecgonine

The presence of a pyridine ring in ecgonine was established by Stoehrs, who obtained α-ethyl-pyridine by distilling the alkaloid with zinc dust. The structural similarity between tropine and ecgonine i.e. their derivation from the same parent substance followed from Einhorn's theory that when anhydro-ecgonine is heated to 280°C with hydrochloric acid it decomposes into carbon dioxide and tropidine.

This result was also deduced from the researches of Liebermann, who converted ecgonine by acid and ecgoninic acid. In this reaction tropinone occurs as an intermediate product. The varying opinions as to the constitution of tropine have therefore also had their reflections on that of ecgonine. Definite proof that the hydroxyl group occupies the same position in ecgonine as in tropine, and that the carboxyl group is attached to a neighboring carbon atom, as indicated in the above formula, was supplied by the work of Willstätter and Müller. It was found that on gentle oxidation with chromic acid ecgonine could be converted into tropine and Ψ-tropine; and further, that the behaviour of ecgonine was not that of α or γ hydroxy acid. Hence the carboxyl and hydroxyl groups must occupy the β- position to one another, and ecgonine is therefore a β-carboxylic acid of tropine. Its degradation to N-methylsuccinimide has already been described before.

When warmed with alkalies, the methiodides of *l, d* and *r*-ecgoninic esters break down into β-cylcoheptatriene-carboxylic acid, as follows:

3-hydroxy-8-iodo-2-(methoxycarbonyl)-8-methyl-8-azoniabicyclo[3.2.1]octane → cyclohepta-1,4,6-triene-1-carboxylic acid + (dimethylamino)iodonium + methanol

Ecgonine possesses the properties of an amino acid, forming salts with bases and acids. The presence of the carbonyl group is not shown by any acid reaction, but is revealed in the stability of the alkali salts towards carbon dioxide, and the production of esters on treatment with alcohols and hydrogen chloride. The alcoholic hydroxyl group may be detected by the formation of esters on treatment with acid chlorides and anhydrides, and in the ease with the compound parts with water and passes into anhydro-ecgonine.

Preparation

l-Ecgonine is obtained by hydrolyzing *l*-coniine – with hydrochloric acid, dilute sulphuric acid, or barium hydroxide. Similarly, the uncrystallable mixture of partly amorphous bases, obtained in quantity as a by product in the isolation of cocaine from coca leaves also yields *l*-ecgonine on hydrolysis.

The preparation of ecgonine from cocaine residues is of value in the technical production of cocaine, since ecgonine can readily be converted into cocaine.

l-Ecgonine crystallizes in monohydric hemimorphic prism (+1mol H_2O) which become anhydrous at 120°C and melt with decomposition at 198°C.

***d*-Ecgonine (*d*-Ψ-ecgonine)** was first obtained by Einhorn by warming ordinary *l*-ecgonine with concentrated alkali. It also results from the treatment of cocaine benzoyl ecgonine, or the alkaloid accompanying cocaine with potassium hydroxide, when the *l*-ecgonine first formed undergoes molecular rearrangement. Liebermann and Giesel obtained it as a fission product of the *d*-cocaine discovered by them among the coca alkaloids. It forms monoclinic prisms of plates of melting point 264°C.

***r*-Ecgonine,** the racemic compound, was prepared synthetically by Willstätter in the following manner, thus completing the synthesis of cocaine.

Sodium tropinone, suspended in either unites carbon dioxide at room temperature unites carbon dioxide at room temperature to give sodium tropinone carboxylate. This compound can be obtained more rapidly by the simultaneously action of sodium and carbon dioxide on the ammo-ketone.

When the crude tropinone carboxylate is reduced in cold, dilute acid solution; with sodium amalgam it yields a mixture of two isomeric compounds having the composition of ecgonine but of different constitution, viz. Ψ-tropine-O-carboxylic acid and r-ecgonine. Only the latter is of interest in this connection, and its formation is probably die to part of the sodium tropinone reacting as the ketonic salt (acetic acid ester) and thus yielding tropinone-β-carboxylic acid, which on reduction passes into r-ecgonine as shown later.

A further synthesis of ecgonine was carried out by Willstätter in connection with the synthesis of tropinone as described before, tropinone carboxylic ester being converted by reducing agents into the ester of r-ecgonine. r-ecgonine contains four symmetric carbon atoms and is not affected by heating with alkalies. It crystallizes with 3 moles of H_2O and melts with decomposition at 251°.

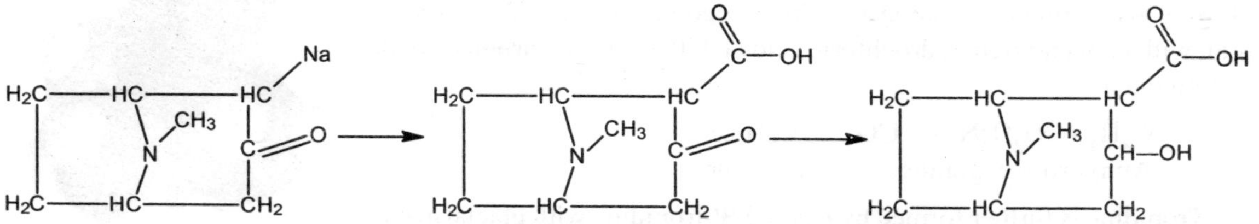

(8-methyl-3-oxo-8-azabicyclo [3.2.1]oct-2-yl)sodium → 8-methyl-3-oxo-8-azabicyclo [3.2.1]octane-2-carboxylic acid → 3-hydroxy-8-methyl-8-azabicyclo [3.2.1]octane-2-carboxylic acid Ecgonine

α - ECGONINE 3-HYDROXYTROPANE 3-CARBOXYLIC ACID

Introduction and Preparation

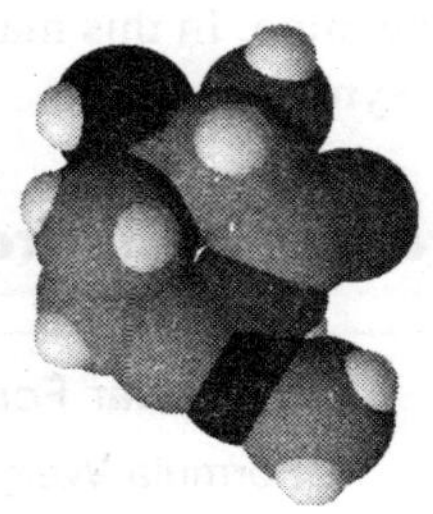

Before the constitution of *l*-ecgonine was known in detail, an attempt had been made to synthesize the racemic compound from tropinone. As a ketone, the latter unites with hydrogen cyanide to form tropinone-cynohydrin. This on hydrolysis yields a compound of the composition of ecgonine, from which it differs in having the carboxy and hydroxyl groups both attached to the same carbon atom. For this structural l- isomeride of ecgonine Willstätter proposed the same ecgonine.

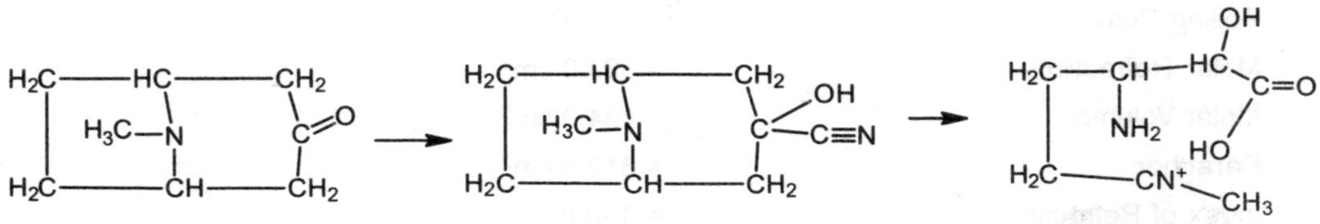

Tropinone → Tropinone cyanohydrin → α-Ecgonine

On benzoylation the methyl ester of α-cocaine. The only interest of these α-compounds lies in the identical with ordinary ecgonine, it followed that the carboxyl group in the latter could not occupy the α-position.

In chemical behaviour α-ecgonine and its derivatives differ remarkably from ordinary ecgonine, where as the methiodides of the ecgonine group, being α-betaines, are easily decomposed by alkalies, those of the α-ecgonine groups being Ψ-betaines, exhibit great stability.

TROPIDINE

Introduction

Tropidine has been mentioned frequently in the forgoing pages. It was first prepared by heating it into 180°C with fuming hydrochloric acid and glacial acetic acid, or with sulphuric acid (Ladenburg). As Einhorn has shown it also formed by heating anhydrous ecgonine (tropene-2-carboxylic acid, with concentrated hydrochloric acid at 280°C, when carbon dioxide is eliminated.

$$C_8H_{12}(CO_2H)N \rightarrow CO_2 + C_8H_{13}N$$

Anhydro – ecgonine Tropidine

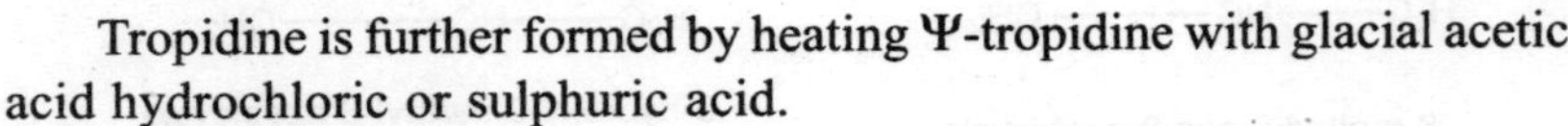

Tropidine is further formed by heating Ψ-tropidine with glacial acetic acid hydrochloric or sulphuric acid.

Cocaine → ecgonine → anhydro ecgonine → tropidine

Atropine → tropine € tropidine

The synthesis of tropidine carried out by Willstätter have already been described. These are of great importance, since tropidine may be converted through Ψ-tropidine into tropine and also into r-ecgonine. In this manner several alkaloids of tropane series particularly atropine are accessible from the synthetic side.

Properties of Tropidine

Molecular Formula	$= C_8H_{15}N$
Formula Weight	= 125.2114
Composition	= C(76.74%) H(12.07%) N(11.19%)
Boiling Point	= 162°C
Molar Refractivity	$= 38.60\ cm^3$
Molar Volume	$= 134.0\ cm^3$
Parachor	$= 313.5\ cm^3$
Index of Refraction	= 1.488
Surface Tension	= 29.9 dyne/cm
Density	$= 0.934\ g/cm^3$
Polarizability	$= 15.30\ 10^{-24} cm^3$
Monoisotopic Mass	= 125.120449 Da
Nominal Mass	= 125 Da
Average Mass	= 125.2114 Da

Chemical Properties

Tropidine is a liquid base with a odour like coniine. It is less soluble in hot water than cold. The aqueous solution turns litmus paper blue.

When treated with excess to bromine at 170°C to 180°C, tropidine yields ethylene dibromide and di-bromo pyridine. On exhaustive methylation, tropidine first gives α-methyltropidine and finally, by distillation of α-methyl tropidine-methylammonium hydroxide it yields tropilidene or cycloheptatriene.

$$C_7H_9N(CH_3)_3OH \rightarrow C_7H_8 + N(CH_3)_3 + H_2O$$

Alakloids of the Tropane Series

To this group belong the alkaloids of the coca alkaloids

Alakloids of the Solance

In many members of the Solanum family, such as *Atropa belladonna* (Deadly Nightshade) *Datura stramonium* (Thorn apple) *Hyocyamus niger* are found a number of alkaloids closely related to each other in their properties and chemical constitution. Chief among these two isomers of the composition $C_{17}H_{23}NO_3$, viz optically inactive atropine and lævorotatory, hydroxylamine.

Atropine is actually the racemic modification of hyoscyamine.

ATROPINE

History and introduction

Mandragora (mandrake) was described by Theophrastus in the fourth century B.C. for treatment of wounds, gout, and sleeplessness, and as a love potion. By the first century A.D. Dioscorides recognized wine of mandrake as an anæsthetic for treatment of pain or sleeplessness, to be given prior to surgery or cautery. The use of Solanaceæ containing tropane alkaloids for anæsthesia, often in combination with opium, persisted throughout the Roman and Islamic Empires and continued in Europe until superseded by the use of ether, chloroform, and other modern anaesthetics.

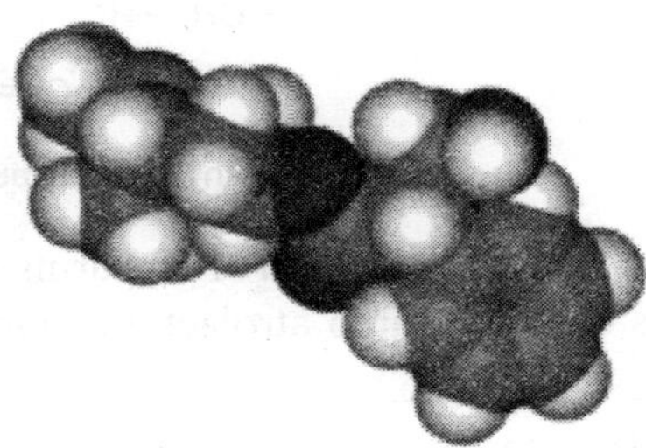

Atropine extracts from the Egyptian henbane were used by Cleopatra in the last century B.C. to dilate her pupils, in the hope that she would appear more alluring. In the Renaissance, women used the juice of the berries of *Atropa belladonna* to enlarge the pupils of their eyes, for cosmetic reasons; "bella donna" is Italian for "beautiful lady".

The mydriatic* effects of atropine were studied among others by the German chemist Friedrich Ferdinand Runge (1795-1867). In 1831 the pharmacist Mein succeeded the pure crystalline isolation of atropine. The substance was first synthesized by German chemist Richard Willstätter in 1901.

Atropine is a tropane alkaloid extracted from the deadly nightshade (*Atropa belladonna*) and other plants of the family Solanaceae. It is a secondary metabolite of these plants and serves as a drug with a wide variety of effects. It is a competitive antagonist for the muscarinic acetylcholine receptor. Being potentially deadly, it derives its name from Atropos, one of the three Fates who, according to Greek mythology, chose how a person was to die.

Preparation and Constitution

In 1863 Kraut discovered that atropine, on being with aqueous baryta, decomposed to yield tropine and atropic acid. A year latter it was shown by Lossen that the primary product of decomposition was not atropic acid, $C_9H_8O_2$, but tropionic acid $C_9H_{10}O_3$ and that the latter was then converted into the former by loss of 1 mol water, consequently the change undergone by atropine is mearly the hydrolysis of an ester into acid and (basic) alcohol.

$$\underset{\text{Atropine}}{C_{17}H_{23}ON} + H_2O \rightarrow \underset{\text{Tropine}}{C_8H_{15}NO} + \underset{\text{Tropic acid}}{C_9H_{10}O_3}$$

By reversing the above process Ladenburg in 1879, affected a partial synthesis of atropine.

On treating tropine tropate with hydrochloric acid, he succeeded in regenerating atropine, thus providing it to be tropine, ester of tropic acid.

The problem of the constitution of atropine therefore resolved itself into two parts the study of tropic acid and that of tropic.

The structure of tropic acid was soon cleared up and the compound synthesized by Ladenburg and Rüghiemer. It is however; more readily prepared by the synthesis of E. Müller in which Phenylacetic ester $C_6H_5CH_2COOC_2H_5$ is condensed with formic acid to give formyl-phenyl acetic ester.

ethyl 2-formylbutanoate → ethyl 2-(hydroxymethyl)butanoate

Another method of preparing this important acid is that due to McRenzie and Wood. Acetophenone is converted into atrolactinic acid which on being heated under reduced pressure gives atropic acid.

*Dilation of cornea.

The latter unites with hydrogen chloride in ethereal solution to form β - chlorohydratropic acid, which finally yields tropic acid on being boiled with aqueous sodium carbonate.

KCN

benzophenone

2-hydroxy-2-phenylpropanenitrile

Atrolactinic acid

The presence of an asymmetric carbon atom in tropic acid indicated the possibility of resolving the acid into its active components and thus of preparing optically active atropines. The resolution of tropic acid was affected by Ladenburg by means of the quinine salt and from the active components the *active atropines* were then built up. By using other acids in place of tropic acid, Ladenburg prepared other esters of tropine, to which he gave the collective name tropeiines. These artificial alkaloids are being described below.

It was not until much latter that the structure of the alcohol tropine the remaining hydrolysis product of atropine, was successfully elucidated and its synthesis accomplished. This has been described in detail previously.

Hence the complete **synthesis of atropine** involves the following stages: (*i*) Synthesis of glycerol, (*ii*) Glycerol into Glutaric acid, (*iii*) Glutaric acid into suberone, (*iv*) Suberone into tropidine, (*v*) tropidine into tropine, (*vi*) Synthesis of tropic acid, and (*vii*) Atropine from tropine and tropic acid.

HOMATROPINE

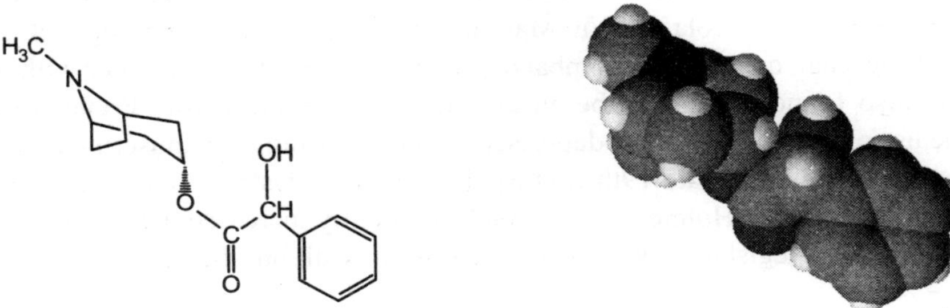

This is also called phenyl-glycolyl tropine it is by virtue of its physiological action the most important compound of the tropine group after atropine and hyoscyamine. It is prepared from tropine and mandelic acid and crystallizes from absolute either in transparent prisms, which are deliquescent and melts at 95°C, it is also found in its hydro bromide state, and in this state is possesses an almost powerful effect of dilating the pupil of the eye like the atropine but the effect disappears comparatively rapidly. It behaves similarly with paralysis of the power accommodation. Homatropine is a much weaker poison than atropine and therefore used in the eye surgery.

According to the investigations of Ladenburg and Völkers the mydriatic power is not solely dependant on the presence of the tropine residue in the molecule it is also necessary for the latter to be united to an containing an hydroxyl grouping.

THE COCA ALKALOIDS

The leaves of the Erytroxylon coca contain a large number of alkaloids. In addition to the hygrine, already described among them Cocaine, Tropacocaine are striking ones.

All these compounds are tropane derivatives and therefore contain pyrrolidine nucleus, with the expectation of tropacocaine they all yield ecgonine as the basic product of decomposition, and stand in close relationship to the Solanaceæ alkaloids.

Before reading the following description the explanation of ecgonine must be read carefully. Among all the coca alkaloids only *l*-cocaine is used for therapeutic purpose but the other being not used for physiological purpose but used in the preparation of *l*-ecgonine.

COCAINE

History

Originally consumed without any processing, the chewing of coca leaves was popular among South American natives long before the arrival of the Spanish in the 16th century. The leaves were chewed in a manner consistent with modern use of coffee, chewed for a small burst of energy or stamina. The Spanish explorers noticed how the natives used the coca leaves and themselves partook in some cases, but the practice of chewing the raw leaves did not become especially popular among Europeans. Coca's turning point in Europe came in 1860 when Albert Niemann extracted pure cocaine powder from coca leaves. This refinement allowed the use of cocaine in many different medicinal products and beverages, most notably Coca-Cola and Vin Mariani. Freud began experimenting with cocaine around this time, consuming small quantities to combat depression, sharing his experience with other European physicians who also found cocaine to be an effective topical anesthetic. Freud became a fervent supporter of the use of cocaine as an anti-depressant, even publishing a manuscript detailing its virtues. Conan Doyle stood alone in the late 19th century depicting the destructive qualities of cocaine in his consulting detective, Sherlock Holmes. As cocaine's popularity increased, health risks were noted and seized upon by American legislators, who made the substance all but illegal in 1916.

Pharmacology

Cocaine in its purest form is a white, pearly product. Cocaine appearing in powder form is a salt, typically cocaine hydrochloride. Street market cocaine is frequently adulterated or "cut" with various powdery fillers to increase its weight; the substances most commonly used in this process are baking soda; sugars, such as lactose, dextrose, inositol, and mannitol; and local anesthetics, such as lidocaine or benzocaine, which mimic or add to cocaine's numbing effect on mucous membranes. Cocaine may also be "cut" with other stimulants such as methamphetamine.

Preparation of Cocaine

> This is the manufacturing process of cocaine intended only for educational purpose, it is not by any means that author is indulging the drug trafficing or drug addiction, the cocaine is a organic chemical substance and therefore has been included in detail students are requested not to use this process for any bad and dirty purpose but take it as a source of knowledge and behave like a matured chemist.

The leaves of Erythryoxylon coca, from which cocaine is obtained, were used by the Incas at least 1,000 years ago. The annual consumption of coca leaf in South America (mainly Bolivia and Peru) by about twenty million users amounts to about 150,000 pounds of cocaine. They chew their leaves with lime (CaO, etc.) which degrades cocaine to ecgonine - but this compound still relieves hunger and fatigue. Cocaine produces little effect orally, so sniffing is the preferred route of administration. Regarding its supposed aphrodisiac properties, it is interesting that one native legend ascribed the magical origin of the coca leaf to the fascination of a conqueror for the bright eyes of a princess who loved him to death. The plant grows in Australia and Africa, as well as in Western South America, and much cocaine was produced in Java. Other species of the genus seem to produce little or no cocaine. Coca-cola got its name from the coca leaf extract which it contained (as did a variety of wines) until 1904. Neither tolerance nor physical addiction to cocaine seem to occur, so sniffing it occasionally should be quite safe.

Cocaine has structural and pharmacological similarities to the active constituents of belladonna and jimson weed and can likewise space you out in very undesirable ways, expecially if used frequently. Also, as with every drug, some people are very sensitive to it and can become paranoid, etc. with very little exposure. Cocaine base ("free base") is much more euphorigenic than cocaine and consequently much more damaging. It seems to have the addictive pull of heroin for many users and is probably best avoided. Coca leaf contains *l*-ecgonine, cinnamyl cocaine and alpha and beta-truxillyl ecgonine (cocamine), which can be converted to cocaine, but d-ecgonine or pseudoecgonine leads to isomers which are devoid of the strong central stimulating effects of *l*-cocaine. During the process of isolation from the leaf, *l*-cocaine is converted to ecgonine (tropan-3-beta-of-2-beta-carboxylic acid), which is easily reconverted to *l*-cocaine (benzoylecgonine methyl ester).

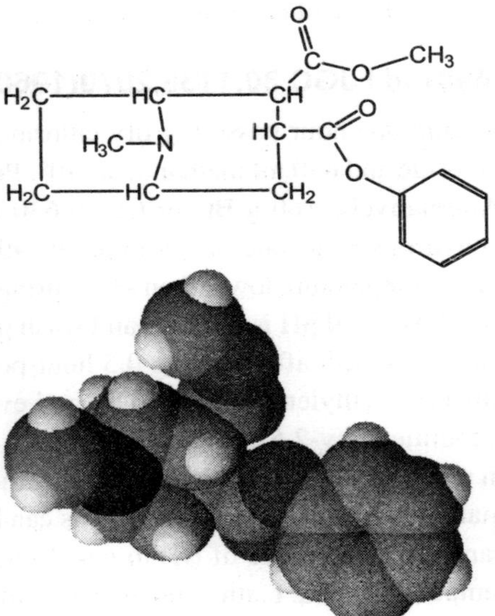

Cocaine Extraction

Cocaine can be extracted from the leaves with almost any organic solvent. Moisten the dried, powdered leaves with Na carbonate solution and extract with cold benzene or petroleum ether. Extract the organic solution with small amounts of dilute sulfuric acid and basify the extract with Na carbonate (the alkaloids precipitate). Dissolve the precipitate in ether, separate the ether from the aqueous Na carbonate and dry and evaporate in vacuum the ether. Dissolve the residue in methanol and heat with sulfuric acid or methanol HCl; dilute with water and extract with $CHCl_3$. Concentrate and neutralize the aqueous layer and cool to precipitate methylecgonine sulfate, which is converted to cocaine in one step. The alkaloids can also be extracted directly from the powder with dilute sulfuric acid.

Cocaine from Coca Paste

This process is optional since the paste is usually greater than 70% cocaine.

Dissolve one gram coca paste in 10 ml 3% sulfuric acid, cool to 0°C and add with stirring 8 ml 6% $KMnO_4$ and 10% sulfuric acid, one ml at a time over one hour. Let stand 0.5 hour and add powdered oxalic acid with stirring until the precipitate which has formed dissolves. Extract twice with ether, basify the aqueous solution with NH_4OH and extract four times with 18 ml ether. Dry and evaporate the ether in vacuum to get cocaine. The aqueous solution contains ecgonine, which can be converted to cocaine as shown below.

Cocaine Synthesis

The zinc-mercury amalgam used for reduction here and elsewhere can be prepared as follows. Add mossy zinc (powdered will probably do) to 1 % aqueous $HgCl_2$, stir awhile, pour off the water and use the Zn-Hg residue for reduction.

Method I JGC 30,1138,2070(1960); CA 53,423(1959)

During one hour pass 70 g dry chlorine gas into a stirred solution of 68 g furan in 630 ml methylene chloride and 630 ml methanol at -40°. Protect from moisture ($CaCl_2$ drying tube) and use dry methanol. Alternatively, 160 g Br_2 or Cl_2 in 630 ml methylene chloride (or methanol in which case add 200 g anhydrous K acetate or pyridine) at -40° is added dropwise to 68 g furan in 630 ml methanol at -40° (keep temperature lower than -30° during the reaction). Stir 0.5 hour and pass in dry NH_3 (or concentrated NH_4OH) until pH is 8. Filter and wash precipitate with 3 × 50 ml methylene chloride. If only methanol has been used, after stirring 0.5 hour pour into 3 L cooled, aqueous, saturated $CaCl_2$ and extract with ether or methylene chloride. Dry and evaporate in vacuum the combined organic solutions to get 90 g 2,5-dimethoxy-2,5 dihydrofuran (I) as a colorless liquid (can distill 40/4 or 71/17). Ethanol can be used in place of methanol, but add 190 ml ether (if reaction temperature rises too high, the methyldiacetyl of maleindialdehyde is produced - this can be reduced to the methyldiacetyl of succinic dialdehyde, which can be used in place of (II) to get (IV)). Hydrogenate 47.5 g (I) in presence of 5 g Raney-Ni at room temperature and 1 atmosphere H_2 with stirring ($NaBH_4$-Ni reduction described at start should also work). After absorption of 7.2 L H_2 over two to three hours, filter and wash catalyst with 15 ml ethanol and evaporate in vacuum (or distill 77/20 for ethoxy compound, 53/22 for methoxy compound) to get 40 g 2,5-diethoxy (or dimethoxy)-tetrahydrofuran (II).To a mixture of 360 g 50% KOH and 138 ml methanol, add with stirring at -5° 70.5 g dimethyl ester of acetone dicarboxylic acid (dimethyl-beta-ketoglutarate - see method 3 for preparation) and let temperature rise to about 25° over 0.5 hour. Let stand ten minutes, cool to 0° and add 65 ml ether. Filter, wash precipitate with 65 ml ethanol and 150 ml ether at 0° to get 75 g (III). To 322 ml 1N HCl at 80°, add 41.1 g (II) and stir twenty minutes; cool to 10°, add 211 ml 1N HCl, 98.2 g (III), 26.4 g Na acetate and 28.2 g methylamine HCl. Stir four hours at room temperature, cool to 10°, and saturate with 410 g KOH. Extract four times with methyl-Cl or benzene (75 ml each, fifteen minutes stirring) and evaporate in vacuum to get the methyl ester of tropan-3-one-2-COOH (IV), which precipitates from the oil (can distill 85/0.2). Test for activity. Dissolve 28.3 g (IV) in 170 ml 10% sulfuric acid; cool to -5° and treat with 3.63 kg 1.5% Na-Hg amalgam with vigorous stirring at 0°. See below for easier methods of reducing (IV). Keep pH about 3.2 by adding 30% sulfuric acid and continue stirring 0.5 hour or until 3 drops of the mixture fail to give a red color with a 10% solution of $FeCl_3$. Filter and saturate the solution with 235 g KOH or K carbonate below 15°. Extract with 5 x 250 ml $CHCl_3$ and dry, evaporate in vacuum to get an oil. (The inactive isomer can be separated at this point (if desired) by letting stand five days at 0° (methyl ester of racemic pseudo-ecgonine precipitates)). Mix the oil plus any precipitate with an equal volume of ether and filter. Add 250 ml dry ether until no more precipitate forms, then filter (test precipitate for activity - if active, this step is unnecessary) and stir with activated carbon 0.5 hour. Filter and evaporate in vacuum, dissolve the brown liquid in 17 ml methanol, and neutralize with 10% HCl in dry ether. Evaporate the ether until the two layers disappear and let stand two hours at 0° to precipitate the racemic methylecgonine (V). Filter, wash with 1: 1 methanol:ether at 0°. Can purify by dissolving in methanol and washing with 1:1 methanol:ether and ether. To prepare the Na amalgam, use an electrolyzer with an Hg cathode, Ni anode, and 40% NaOH solution; current about 29 amps, 7.5 volts. Heat 9.35 g (V) on water bath ten hours with 18.7 g benzoyl-Cl and pour the liquid formed into 250 ml ether; evaporate in vacuum. The powder formed on rubbing the residue is dissolved in 85 ml ice water and neutralized with 20% NH_4OH. Racemic cocaine (VI) is filtered off, washed with 12 ml ice water and dried over $CaCl_2$. To get the HCl of (VI), dissolve it in seven times its weight of ether containing HCl in ethanol and wash the precipitate with 1:3 methanol:ether, then ether.Alternatively, mixture of 4 g (V), 36 ml benzene, 1.6 g Na carbonate and 7 ml benzoyl-Cl; stir and heat 96-100° for ten hours. Evaporate in vacuum, cool to 0° and add 40 ml ice water. Acidify with HCl to pH 5 and extract with 3 x 20 ml ether. Neutralize the aqueous solution with 20% NH4OH to separate an oil from which cocaine precipitates on standing (from JGC 30,3228(1960)).For an electrolytic method of producing (I). For other methods of synthesizing (II). For the reduction of the tropinone (IV) to ecgonine (V), lithium aluminum hydride or $NaBH_4$ give about 50%, the Na-Hg method described above about 40%, and Al triisopropoxide about 25% of the inactive pseudoecgonine. The latter method, which appears to be the best, involves heating at 82° 1.5 hours in isopropanol with Al triisopropoxide (Chem. and Ind. 664(1957)). For NaBH4, reflux six hours in methanol. A method said to be superior to that given above for the conversion of methylecgonine to cocaine follows.

Cocaine from Ecgonine BER 89,679(1956)

Dissolve 8.66 g ecgonine in 100 ml methanol and bubble dry HCl gas through for 0.5 hour. Let stand two hours at room temperature and then reflux gently for 0.5 hour. Evaporate in vacuum, basify with NaOH and filter to get 8.4 g methylecgonine (V) (recrystallize isopropanol). Add 4.16 g (V) and 5.7 g benzoic anhydride in 150 ml benzene with $CaCl_2$ tube to exclude water and gently reflux four hours.

Cool in ice bath, acidify with HCl and dry, evaporate in vacuum (or extract with ether, basify with NaOH, saturate with K carbonate and extract with $CHCl_3$:dry and evaporate in vacuum) to get 6 g red oil which precipitates (VI) with addition of a little isopropanol.

Method 2 JOC 22,1391(1957)

Suspend 40 g beta-ketoglutaric acid in a mixture of 60 ml glacial acetic acid, 43 ml acetic anhydride and stir three hours at 10°. Filter, wash precipitate with benzene and dry in vacuum over KOH about two hours to get 30 g beta-ketoglutaric anhydride (I). Dissolve 13.5 g (I) in 50 ml cold methanol and let stand one hour at room temperature. Add this solution to a solution of 10 g methylamine HCl, 4 g NaOH in 850 ml water and stir in 125 ml 0.8N succindialdehyde (preparation below). Let stand twenty-four hours at room temperature, take pH to 4 with 6N HCl and wash with 35 ml $CHCl_3$. Dry and evaporate in vacuum (or basify with 20 ml 4N NaOH and 4 g $KHCO_3$, extract with 9 x 100 ml $CHCl_3$ and dry, evaporate 75% of the $CHCl_3$ on steam bath, then evaporate in vacuum) to get 16 g yellow oily 2-carbomethoxytropinone (methyl tropan-3-one-2-carboxylate). Recrystallize by dissolving the oil in 30 ml hot methyl acetate and add 4 ml cold water and 4 ml acetone; let stand three hours at 0°, filter and wash precipitate with cold methyl acetate. This product is identical with (IV) of method 1 and is converted to cocaine as already described.

Succindialdehyde JOC 22,1390(1957)

Suspend 23.2 g succinaldoxime powder in 410 ml 1 N sulfuric acid and add dropwise with stirring at about 0° a solution of 27.6 g Na nitrite in 250 ml water over three hours. Stir two hours at room temperature (keeping air out), stir in 5 g Ba carbonate and filter. The succindialdehyde should then be extracted from the basic solution with ether and the ether dried and evaporated in vacuum. For succinaldoxime preparation.

Method 3 JOC 22,1389(1957)

Add with stirring over 1.5 hours 192 g powdered anhydrous citric acid in 32 g portions to 202 ml (383 g) fuming sulfuric acid (21%). Make the first two additions at 0° carefully; the other four at 15°. Stir one hour at room temperature, and for three hours at 35° and 17 hours at 25°. Add dropwise with stirring below 0°, 500 ml methanol over three hours. Keep about fifteen hours at room temperature and add to a stirred mixture of 700 g $NaHCO_3$, 500 g ice and 200 ml water. Filter, wash precipitate with 150 ml 50% aqueous methanol and extract the filtrate with 7 x 400 ml ether. Dry and evaporate in vacuum (can distill 85/1) to get 110 g oily dimethyl-beta-ketoglutarate (I). Use this in method 1 or as follows. Dissolve 33.6 g KOH in 150 ml methanol and add dropwise at 0° over 0.5 hour (or at room temperature over one hour) to 43.5 ml (I) in 10 ml methanol. Let stand three hours at room temperature, add 50 ml ether and refrigerate twelve hours to precipitate the dipotassium salt of monomethyl-beta-ketoglutarate (II). Dissolve 10 g succindialdehyde in 200 ml water at -5° and add 41 g (II) and 11.8 g methylamine HCl. Let stand a few hours at room temperature and proceed as in method 2.

Method 4 JOC 22,1389(1957)

Mixture of 1.35 g Na methoxide (Na in methanol), 3.48 g tropinone (which can be obtained by K dichromate oxidation of tropine), 4 ml dimethylcarbonate and 10 ml toluene. Reflux 0.5 hour, cool to 0°

and add 15 ml water containing 2.5g NH4Cl. Extract with 4 x 50 ml $CHCl_3$, dry and evaporate in vacuum and dissolve the oil in 100 ml ether. Wash twice with a mixture of 6 ml saturated aqueous K carbonate and 3 ml 3N KOH (dry and evaporate in vacuum the ether to recover unreacted tropinone). Take up the oil which separates in saturated aqueous NH_4Cl and extract it with $CHCl_3$. Dry and evaporate in vacuum to get an oil which is dissolved in hot acetone. Cool, add a little water and rub to start precipitation of 1.5 g 2-carbomethoxytropinone. This is identical with (IV) of method 1, and can be recrystallized and converted to cocaine as already described.

Cocaine Manufacture and the Cocaine Traffic

The coca plant is an evergreen, native to South America, particularly the countries of Peru, Bolivia, Brazil, Chile and Columbia, and should not be confused with the cocoa plant, from which chocolate is made. Although the coca plant is native to South America, it has been successfully cultivated in Java, West Indies, India and Australia.

The coca plant is grown on mountain slopes or terrace uplands that have a tropical or semi-tropical climate. Actually, the plant is grown under conditions which are little suited for other crops. These mountainous areas of South America vary in altitude from 1,000 to 6,000 feet above sea level and temperatures of 68 to 86 degrees Fahrenheit. The most suitable conditions for the development of the coca plant are clay type soils, rich in humus and iron content, situated in sheltered upland valleys and exposed to constant humidity and rain precipitation. Under ideal conditions, the plant can survive, for a century or more, growing steadily in strength. In the cultivation of the coca plant, the seeds are usually taken from a plant three or more years old, placed in containers and germinated in damp sheltered nurseries. They are watered heavily for five days until they begin to swell after which they are planted in a mixture of humus sand and earth, in equal proportions, shaded and abundantly watered. After about a week and a half, the shoots begin to appear and the germinated seeds can be transplanted within 2 months. The sparsely leaved plant is usually 6 to 10 inches in height and is transplanted in the open since it has become resistant to most climatic variations. The young plant is usually planted in small trenches varying in density from 1 to 4 plants per square yard.

Once the young plant has been transplanted (usually in the wintertime) into open fields, there is very little the cultivator does except leave the plants to themselves. Where dampness is constant and rain is regular, even irrigation is unnecessary. After approximately one year from the transplanting, the coca plant yields its first crop of leaves which is generally the reason the coca bush has been cultivated in the first place. The plant normally yields 4 crops of leaves per year. The coca plant is a shrub-like bush which grows from 5 to 10 feet tall with widely branched trunks containing twigs which become densely populated with leaves toward the ends. The green smooth edged leaves vary from 1 to 3 inches long and smell very similar to tea leaves. Normally, in order for the harvesting of the coca leaf to be profitable, there must be a minimum of 72,000 plants for every 10,000 square meters and the plant must last for over 30 years. Approximately 10 million coca plants produce 700,000 kilograms of coca leaf. The average coca leaf contains from 0.5 to 1% of the alkaloid cocaine although there are various factors influencing the cocaine content including atmospheric conditions, age and condition of the plant, quality of the soil, fertilizers used, timing of cultivation and harvesting, the drying process, etc. It is estimated that one man can harvest approximately 30 kilograms of leaves in a day. The drying process is very important and takes approximately 2 days of at least 3 hours of daily sunshine. During the process, the leaves must be turned over for even drying. If the drying is to extensive, the leaves will become too dry and lose their commercial value. In the drying process, the coca leaf loses more than 75% of its original weight. The leaf is divided into 3 basic categories:

1. Dark green coloured leaf, dried by mechanical means or by airing and pressed into bales. This form is best suited for use in export.
2. Dark coloured leaf resulting from defective drying and deliberately beat in order to meet wide demand for consumption.
3. Leaves which as a result of neglect, dampness, delay in drying or disease, have lost some of their alkaloid content, and are of practically no commercial value. This type of coca leaf is often used as a local fertilizer.

Once the leaves are dry, they are pressed and wrapped into packages of 50 or 30 kilograms. In 1965, 1 kilogram of dried coca leaf, in Bolivia, was valued at approximately one dollar.The coca leaf is commonly chewed by the natives of South America. The natives claim that the cocaine depresses their hunger and increases their strength. The leaves are very bitter when chewed and are often flavored with another substance such as lime. It has been estimated that over 90% of the Indians chew the coca leaf. The native chews, on an average, about two ounces of coca leaf daily and is often characterized by blackish red deposits on his teeth.The coca leaf is either consumed by the natives of South America or exported to other countries for consumption. Another use of the coca leaf is in the extraction of cocaine either for illegitimate or legitimate use. The majority of the legal and/or clandestine cocaine factories are in South America due to the cost and bulk of transporting the whole leaf. In 1961, Bolivia produced an annual crop of from 12,000 to 18,000 tons of leaves although only half reached the legal market. The alkaloid cocaine is extracted from the coca leaf in basically three different chemical procedures. These procedures are used both in licit and illicit labs in the production of cocaine.

Manufacture of Cocaine

According to a chemist who assisted in the legal manufacture of cocaine, there are three basic methods of extracting cocaine from the coca leaf:

Coca Leaves
↓
Egonine
↓
Cocuine

1. The dried coca leaf is treated, through a chemical process, with an acid solution such as sulfuric acid, producing raw cocaine or coca paste. The coca paste which contains approximately 70% cocaine, is put through another chemical process with hydrochloric acid creating a hydrochloric salt or cocaine hydrochloride which is soluble in water. This particular process is very time consuming and can take from 1 to 2 weeks to complete. This process is used by both the legitimate and illicit manufacturers of cocaine.

2. The dry coca leaf is treated in a chemical process with a basic solution such as sodium carbonate, producing raw cocaine. The raw cocaine is then put through another chemical process with hydrochloric acid creating a hydrochloric salt or cocaine hydrochloride. This process is less time consuming than process No. 1 and it is probably the one preferred by the illicit manufacturer.

3. The third process is more advanced and technical than the other two procedures. The basic advantage of this process is that it gives a greater yield. The dried coca leaf, with it's various alkaloids including cocaine, is broken down into ecgonine which is the chemical base or core of the cocaine molecule. The ecgonine is then treated with methyl iodide and benzoic anhydride in a chemical process creating pure cocaine.

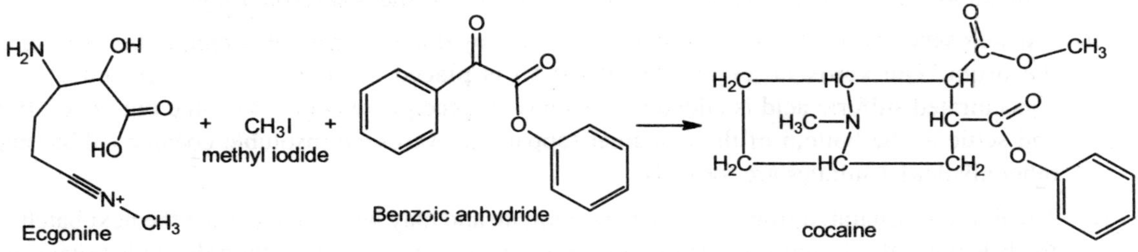

Ecgonine + CH_3I methyl iodide + Benzoic anhydride → cocaine

Cocaine Preparation

The Peruvian coca leaves, because of their richness, are commonly used in the extraction process as described in 1 or 2. When the dried coca leaves have a low cocaine content, the ecgonine process is preferred. Normally, it takes approximately 100 pounds of dried leaves to produce one pound of cocaine.

A chemist from the Federal Bureau of Narcotics and Dangerous Drugs, who was in Bolivia to observe clandestine cocaine operations, related the following step-by-step procedure for manufacturing cocaine. The method can be conveniently divided into three major steps: (1) extraction of cocaine from the leaf and chemical conversion to the sulfate; (2) treatment of cocaine sulfate with potassium permanganate and conversion to the free base (aka paste); and (3) conversion of the paste or free base to cocaine hydrochloride. In general, steps (1) and (2) are carried out in "sulfate" labs while step (3) is performed in "crystal" labs.

Coca Leaves to Cocaine Sulphate

Step 1:

(*a*) A sufficient volume of warm water is used to dissolve 26 kgm of potassium carbonate. This solution is poured into a "cut-out" drum containing 250 pounds of dried coca leaves (the volume of solution is just enough to cover the leaves).

(*b*) The coca leaves and carbonate solution are treated in one of three ways:

1. The mixture is stirred by hand.
2. The mixture is stepped on by the local Indians.
3. The mixture remains untouched (the first two methods generally take one day if done by strong individuals while the third method involves a period of four days; the treatment of the leaves with an aqueous solution of potassium carbonate allows penetration of the solution into the leaves converting any cocaine salts to the free base, allowing for subsequent extraction into kerosene).

(*c*) To the potassium carbonate solution and the leaves are added 200-400 L kerosene. The result of this is a greenish viscous liquid of about 300-400 L. Water and kerosene are immiscible but apparently "bridging" materials are extracted from the leaf which allows miscibility between the potassium carbonate solution and the kerosene; it was also explained that what we know as kerosene is also called benzin and paraffin, depending who is manufacturing the cocaine; this explanation clarified pre-existing conflicts about certain method terminology.

(*d*) The kerosene extract is separated from the leaves by draining through a plug in the bottom of the drum. The kerosene extract (300-400 L) is placed in another container and 1 L of concentrated sulfuric acid is added very slowly. A precipitate begins to come out of solution and settle to the bottom of the container (apparently kerosene-insoluble cocaine sulfate and other alkaloidal sulfates are formed).

(*e*) The liquid is separated from the sulfate precipitate and may be used to extract the next batch of fresh leaves (the liquid would probably have to be reconstituted, though, with potassium carbonate, etc.).

(*f*) The sulfate residue, about 1 kgm, is allowed to dry in the sun for about one day.

Cocaine Sulfate to Cocaine Base (paste)

Step 2:

(*a*) Into a container holding 6 L of water and 360 cc of concentrated sulfuric acid the sulfate precipitate from the preceeding step is added and the solution stirred.

(*b*) In another container 1 L of water and 1 kgm of commercial potassium permanganate are mixed.

(*c*) The permanganate solution of (b) is slowly poured into the sulfuric acid solution of (a).

(*d*) The resultant purple-colored solution is then filtered through paper; if the filtrate is colored it is passed back through the filter (the use of potassium permanganate in the manufacturing process has been well-established; it is probably added as a decolorizing agent, with most of the colored residue remaining on the filter paper; commercial potassium permanganate is easily obtained in LaPaz through pharmacies; it is used by the local inhabitants to bathe the feet).

(*e*) To filtrate of about 6 L is added 1 L of ammonia; a precipitate is formed which is dried in the sun or under ultraviolet light (the resultant precipitate is cocaine free base or paste plus impurities).

Cocaine-base (paste) to Cocaine Hydrochloride

Step 3

(*a*) To the approximately 1 kgm of paste in a container is added 10 L of acetone and the resultant solution filtered through paper (the residue on the paper is probably inorganic in nature; ether can be used instead of acetone with some modification in the procedure; both ether and acetone can solubilize cocaine base while they are both good solvents for the crystallization of the hydrochloride salt).

(*b*) To the filtrate is added 10 L more of acetone; the resultant 20 L is passed through new filter paper. The temperature of the acetone should be about 150 C. in order to accept the hydrochloric acid and alcohol.

(*c*) To the 20 L of acetone are added 300 cc of concentrated hydrochloric acid and 300 cc of absolute ethanol; on the addition of absolute ethanol cocaine hydrochloride starts crystallizing out.

(*d*) After 3-4 hours crystal formation is complete and the cocaine hydrochloride crystals are collected on filter paper and dried in air.

Kerosene, which is used in great volumes in the initial extraction procedure is not controlled in Bolivia and is, in fact, used by the populace for heating their homes. It would theoretically be very easy to obtain kerosene in large quantities through pharmacies in La Paz. The inorganic reagents such as sulfuric acid, hydrochloric acid, ammonia, and potassium carbonate are not controlled. Ether and absolute alcohol, on the other hand, are controlled substances. Diluted alcohol (about 50%) is not controlled though and, accordingly, can be converted to absolute alcohol.

Trafficking in Cocaine

A. Legal

Once the cocaine has been legally produced from the coca leaf, it is exported to various countries for medicinal use, basically as a topical local anesthetic (applied to the surface, not injected, only treating a particular area). In the United States the crystalline powder is imported to pharmaceutical companies who process and package the cocaine for medical use. Merck Pharmaceutical Company and Mallinckrodt Chemical Works distribute cocaine in crystalline form (Hydrochloride Salt) in dark colored glass bottles to pharmacies and hospitals throughout the United States. Cocaine, in the alkaloid form (base drug containing no additives such as hydrochloride in the crystalline form) is rarely used for medicinal purposes. Cocaine is still a drug of choice among many physicians as a topical local anesthetic because the drug has vasoconstrictive qualities (shrinks and stops the flow of blood). Synthetic local anesthetics such as novacaine and xylocaine (lidocaine) have also been discovered and used extensively as a local anesthetic.

Dilution or "Cutting" of Cocaine

The paraphernalia and diluting agents for the cutting of cocaine are very similar to those used for heroin. One of the basic differences between "stepping on" (diluting) cocaine as compared to heroin, is that cocaine is usually only diluted down from 20 to 40%. The process for cutting cocaine varies from individual to individual with often times the large dealer using a more elaborate process, but the basic operation is the same throughout cocaine traffic. The basic paraphernalia used is:

Diluting agent, scales, measuring spoons, flat nonporous surface, razor blade, playing card or some other sharp-edged instrument, sifter or nylon stocking, funnel and packaging container such as rubber condoms, tinfoil bindles or plastic baggies.

The cutting or diluting agent used for cocaine again varies with the individual and the substance that is readily available to that individual.

Some of the ways of ascertaining the approximate percentage of cocaine and the cutting agent are:

1. **Quantitative chemical analysis** which is an elaborate process requiring a qualified chemist and some elaborate laboratory equipment.

2. **Cocaine drug testing kits** either manufactured for law enforcement purposes or produced by the underground. These testing kits are simply presumptive color tests. The basic color test used for cocaine is cobalt thiocyanate. The cocaine or any of the other substances from the caine family will form a brilliant blue flaky precipitate in the cobalt thiocyanate. This is an indication that the product is cocaine, procaine, tetracaine, etc. In order to determine whether there is actually any cocaine and not all procaine, stannous chloride is added to the precipitate causing all of the caines except cocaine to dissolve. If the dealer suspects that the cocaine has been cut with another caine, he can then make a partial determination as to how much of the procaine or other caine is contained in the total powder.

3. **Chlorox test**. It is alleged that the dealer can take suspected cocaine and drop it in a vial of clorox. Presumably the cocaine will dissolve completely and procaine will turn a reddish orange color with any other cut trailing to the bottom of the vial as residue.

4. **Water Test**. It is also alleged by the street dealers, that a determination can be made as to how much cut is in the cocaine by placing the powdered substance in a glass of water. The cocaine will dissolve almost immediately leaving the remaining cut which normally will dissolve slower and not as clear.

5. **Burning test**. The powdered cocaine is placed on aluminum foil and held over a low flame or match. The cocaine will burn clear. A sugar cut will darken and burn a dark brown or black therefore the larger the cut, the darker the burn. Crystallized speed or methamphetamine will pop when burned. Salts do not burn and remain as residue (cuts such as procaine or quinine also burn fairly pure although it is alleged that procaine can be detected by a bubbling of the substance before it burns clear).

6. **Methanol test**. Most common cuts do not dissolve in pure alcohol although cocaine does. Unfortunately for the dealer, procaine and methamphetamine also dissolve in pure alcohol. It is imperative that pure methanol be used since any water in the alcohol will tend to dissolve sugar and salt. Methanol can be obtained in most paint supply stores as methalated spirits. The dealer will take two equal amounts of the cocaine substance and place the equal amounts in two teaspoons next to one another. At this time, 0.25 of a teaspoon of pure methanol is added to one of the spoons. The mixture is then stirred and any powder that remains is compared to the original unaltered amount in the second teaspoon to determine the percentage of the cut. If, for example, 20% of the original amount did not dissolve, the substance tested would be no more than 80% pure. If the suspected cut is procaine, the cocaine substance can be added to sodium carbonate solution. This would dissolve all the cocaine leaving just the procaine.

7. **Use test.** Some dealers will test the percentage of cocaine by inhaling (snorting) it into the nostrils. This is probably the best and most common street test in determining the purity of the cocaine. The tester should standardize the amount snorted so that he will have the ability to distinguish. The tester will look for the swiftness of the high and the "freeze" or numbness the substance causes. If the nasal passages burn and the eyes tear, there is a good possibility the cocaine has been cut with speed. Sugar and salt cuts will often times cause a post nasal drip. Excessive sweating and hyperactivity could mean either a speed or quinine cut was used. Excessive diarrhea would denote a laxative type cut such as epsom salts or menita. Speed tends to cause irregular bowel movement. A greater degree of numbness indicates the presence of procaine or other local anesthesia.
8. **Taste test.** Cocaine has a bitter taste and the addition of any cut will tend to alter that taste. A milk sugar cut will sweeten the cocaine although dextrose has a tendency to sweeten the substance more than lactose. Procaine will be bitter to the taste but will tend to numb the gums and tongue quicker and longer than cocaine. Salt has an after taste and epsom salts are a bit more sour in taste and sandy in texture.
9. **Observation test.** Pure cocaine crystals have a shiny almost transparent appearance and even when crushed, will retain the crystalline sparkle. The crystalline sparkle of cocaine will be dulled by most cuts. Dextrose has less dulling effect than lactose although a speed cuf usually dulls the crystals less than most other cuts. Although salts have a crystalline structure, they tend to be duller than the cocaine crystals. An alleged indication of the purity of the cocaine is the tiny rock-like material contained within the total substance. The tiny rocks are allegedly pure cocaine as they come from the manufacturer. The rock or hard substance can be felt by feeling the powdered substance.

Once the dealer has ascertained the purity of the cocaine and/or the cut or diluting agent used, he is then ready to begin the process of "stepping on" the cocaine. Most dealers will dilute a small portion of the cocaine and then re-test it. Most of the dealers claim that they usually only cut the amount of cocaine that will immediately be sold due to the fact that the cuts have a tendency to destroy the stability of cocaine. It is therefore advantageous to the dealer to keep the cocaine sealed in a cool place such as the refrigerator and in an amber or dark-colored jar to retain the strength of the drug as long as possible. Dealers claim that with time, moisture, warmth, air and sunlight tend to decrease the potency of the cocaine.

The process for "stepping on coke" again varies with individuals but the two basic formulas are similar to those of heroin and are as follows:

1 oz. of Lactose added to 1 oz. of 100% Cocaine = 2 oz. of 50% Cocaine

2 oz. of Lactose added to 2 oz. of 50% Cocaine = 4 oz. of 25% Cocaine

2 oz. of Lactose added to 1 oz. of 100% Cocaine = 3 oz. of 33.3% Cocaine

3 oz. of Lactose added to 1 oz. of 100% Cocaine = 4 oz. of 25% Cocaine

4 oz. of Lactose added to 1 oz. of 100% Cocaine = 5 oz. 20% Cocaine

The dealer will measure out the desired amotnt of cocaine,)tbr instance five level teaspoons, and place it in a pile on a flat nonporous surface suchɒs a record albɒun, mirror or glass plate. He will then measure out the desired amount of lactose and ɾoace it in a sepɾcate pile on the same surface. Then, using a playing card, razor blade, knife or any sh. kp edged instru. hent, the dealer chops the cocaine to take out all the lumps so the cocaine is a fairly fiir powder. Theiiocaine is then sifted through a sifter or nylon stocking producing a fine fluffed powdae and removinɡaforeign material from the substance. Once through the sifter, the cocaine usually hasaı little more vailume since it has been fluffed. The cocaine is sifted into a pile and the same process iɘ repeated with ɪɘ e diluting agent. The dealer will then mix the pile of cocaine into the pile of diluting piɡent. Once thi phas been accomplished, he sifts the diluted cocaine through a sifter trying to get the ɪyixture as equalys possible. The dealer may resift the diluted cocaine to assure an equally distributed mlyture. The siftal cocaine is placed in a single pile. At this time, the dealer is ready to place the cocainelanto packages ɜlar sale.

In larger quantities, the cocaine is usually pickaged in eitbir airtight rubber condoms or plastic baggies that are also airtight. A rubber condom wibl usually hold ɴtom 0.5 ounce to 3 ounces of cocaine whereas a plastic baggie can hold up to 6 ouncesl ɒf cocaine. It al pears that most dealers prefer using the plastic baggie since it is believed that synthe:v: hard plastics:and rubber tend to react unfavorably with cocaine in terms of chemical composition:and therefore ɪa time, decrease the potency of the cocaine. Smaller amounts or gram quantities of cn caine are usuɑly packaged in smaller airtight clear plastic bags, paper bindles or tin foil bindles. It :s very uncoɪn ɪon to see cocaine packaged in toy balloons the way heroin is packaged because of ɡe action of th grubber with cocaine.

Once the dealer has diluted the cocaine, he thill measure out the desired amount to be packaged. This can be done by measuring it on a scale.

For ounce quantities, the dealer will measure en a scale apprc eimately 28 grams and then place this powder into a rubber condom or a plastic baggie a The dealer wɛ d then, using a rubber condom, tie a knot in the condom and possibly fold the openy nd back oveny or added protection. When using a plastic baggie, the dealer will get as much of the : ar out as possil ɛ and either fold the baggie over and seal it with scotch tape or use a twister to seal the ɪad of the bagg i. Another method of determining the proper amount of cocaine to be packaged is by usıcg measuring sɪoons. If the dealer is going to use the most common measure of cocaine, that is a "spoen" or approxiɑeately a gram, he will measure out a level half teaspoon and place it in the proper paɑ aging device. ɜVhen using paper bindles, the dealer will place the cocaine on the extreme inside of therpaper bindle, ıɐsually a square piece of paper with 4 inches by four inches being a common size anordrawing the toıo opposite ends together to form a triangle. The dealer will then begin at the base ofine triangle andirold approximately 0.25 inch folds in the paper bindle until there is approximately an inro of unfolded trcangle left at the top. The outer wings of the paper bindle are then folded inside and the u ifolded top of u e bindle is folded and tucked into the wings. When using tin foil, the dealer will put thealesired amoure inside a square piece of tin foil, fold the square over into a rectangle, seal the top witsea small fold alsd then fold the extreme ends into the inside. Once the cocaine has been packaged, it isn ready to sell teneither smaller dealers or to users.

How to Grow Coca Plants

1. Seeds should be planted as lbon as they fa. lfrom the bush. If they dry out, they will die right away. The only way to keeu them for a muximum of about two weeks, is to keep them in moist (not wet) sphagnum a a cool placeiaOften this initiates germination, so they must be watched for rot or prematur germination.mJnder no circumstances should they be kept dry, since even room humidity i too dry.
2. Vermiculite seems to be the best medium fonoca germination, fine grade if possible. Styrofoam cups are OK, but I prefer sm 'll plastic pots,h "diameter, with holes in the bottom. Seeds should be planted no deeper than sae inch. Pots s sould be raised so as not to saturate the medium. Coca, whether as a seedlingcor a mature pltat, never likes to have wet feet. I think it is better to start them in small pots ther than flatss so there is less damage to the root system when they are transplanted. Forgo the hot pad so I think it is completely unnecessary. Seedlings usually come up in 2 to 4 w eks if they are viable.
3. Since most people don't havse enough roor sn their shower stalls for plants, I'd say forget this one, too. Seeds will germina l in any warme lace, even if the humidity is not too great. A better idea is to place your germinction pots in a crrarium with a coarse gravel layer on the bottom. Do not seal over and allow pienty of ventilasion if you choose to place a layer of glass over the terrarium. Any box of this sort will do. If possible, place a Growlux fluorescent fixture, with two 40 W bulbs, over the tglrarium, espec glly after seeds germinate. A common problem at this state is etiolation (too little light) whicntmakes the plantlets weak and very susceptible to damping off, a fungus attac of the tender stems.
4. Water the seeds when the vermiculite staOs to dry out. Once a day is probably too often, unless you live in a very dry partment. Buts the drainage is good and you have plenty of holes in the bottom of the pots, excess water should drain off. Fungal attack is a real problem in a humid atmosphere and anoiter reason for nteeping the plants out of your shower, a basically unhygenic place for plants.
5. Transplanting: plantlets cancemain in vernpculite starting pots until they are about 2-3 inches tall. The growlights should.be about a foots.bove the plants. I do not recommend clay pots at this stage. They dry out tooerast, especiallyen a dry apartment. Even in one day, a fast-drying shock can kill your plants. Its better to moote into plastic pots, but the size should be increased gradually. A big pot is not nt,cessarily goodt,or a small plant, in fact it is not a good idea at all. From styrofoam cups, I sugaest a two-inceapot, then increase 1-2 inches per transplanting.
6. Soil mixture: forget the verticulite from t tw on. It holds too much moisture and makes for saturated, unhealthy soil. I auggest the folawing: 0.25 coarse clean sand, 0.25 perlite, 0.25 sterilized loam, and 0.25 milled peat. If this seems too light, increase loam and peat. Some sterilized organic compost, screened, may also be added for nutrition.
7. Even when the plants are sti in vermiculito feeding with soluble plant food is recommended. They are heavy feeders and every three weeks or more often is not too often to fertilize. When plants are older it is importa it to give them ron in the form of iron chelate, available as a red

powder sold as KEELATE on the West Coast. A yellow powder, not as good, is sold as SEQUESTRENE. This element should be added about every six months, but strictly according to instructions. Soil must be flushed three times after applying the dissolved iron compound to avoid burning roots. Most yellowed or bleached out leaves are caused by iron deficiency, but this also occurs when plants go deciduous. Periodically, the whole coca bush turns yellow and drops its leaves, every one. Most people freak out when this happens, but if it is otherwise a healthy, vigorous plant, then this is normal. After dropping, new flushes soon appear to renew the foliage. This is more likely to happen with Erythroxylum coca than with E. novogranatense.

8. **Transplanting** depends on the size of the plant and how fast it is growing. If you think your plant needs transplanting, look at the holes in the bottom of the pot to see if any roots are present. If so, then the roots have probably filled the pot and it is time. You can also carefully de-pot the plant by tapping upside down on a table edge. Repotting is probably unnecessary unless the roots have encircled the inner periphery of the pot. Again, the size of the pot should be increased gradually for best growth.
9. **Watering**: most city water, is unsuitable for coca. They are calciphobes and don't like heavy salts in the water. Best to use rainwater, melted snow, bottled spring water or distilled water if they are available. Plants should only be watered if the soil dries out. Stick your finger in the soil. If it feels moist, don't water.
10. **Bugs**: coca is amazingly resistant to insects and mites. Mealy bugs are the worst offenders. These may be removed with a forceps or cotton swab dipped in 50-70% alcohol. Keep infested plants in quarantine. Malathion may be used as a last resort, but then leaves cannot be used until the next flush (of leaves).
11. **Light**: warm, sunny exposure indoors. Full sun (through a window) will not hurt plantlets over 3 inches tall. But no full sun outdoors until they are 3 feet tall. If plants are put out in the summer, they should be protected from sun, rain, and wind, until they are large and strong. Put them in a shady place first, under a tree, etc., end gradually move to a sunnier location. Breezes are good for plants and even indoors a fan on low should be directed towards the plants. It makes them stronger.
12. Plants can also be grown entirely under growlights, or a combination of growlights and window light. Most apartments are not sunny enough for strong growth, so especially in winter, give the plants accessory light. Growlux Widespectrum Tubes seem to work well. We use one Growlux and one regular Sylvania Lifeline tube in each fixture. They work very well. The lamps are suspended 6 inches to one foot above larger plants.
13. Careful removal of the older leaves does not harm plants, but they should be strong and healthy to allow this, and probably three years old if grown indoors.
14. Coca does not like extremes of any kind. 50° F. is the lowest permissible temperature, 90° F. the highest. Sudden temperature changes are especially damaging. Likewise, sudden changes in air humidity or soil moisture. *E. novogranatense* tolerates extremes, especially droughts, better than E. coca, which is a much more delicate plant, but the one which produces the most alkaloid.

15. Coca cuttings root very poorly. We have managed to root some *E. novogranatense* cuttings only after six months in perlite with an initial application of Hormodin #1 rooting hormone. It is better to fertilize your flowers and plant seed. Some varieties are self-compatible (self-fertilizing). Others require two plants of different stylar lengths (long styled × short styled) to produce seed. This is routinely accomplished by bees and other insects in the greenhouse during the summer months and can be done with a fine artists brush at home, merely by dusting pollen from flowers on one plant to those on another with opposite stylar form. In California, outdoor cultivation of coca is possible only around San Diego, if there. Trujillo Coca would probably do well there under irrigation and intensive care. Elsewhere, forget it. I do not subscribe to growing it commercially indoors and doubt if the produce would be worthwhile. Greenhouse and apartment grown leaf is very inferior in flavor and potency. Fresh air and sunshine are in order (as with Cannabis).

Physcial Properties

Molecular Formula	= $C_{17}H_{21}NO_4$
Formula Weight	= 303.35294
Composition	= C(67.31%) H(6.98%) N(4.62%) O(21.10%)
Melting Point	= 195 °C
Molar Refractivity	= 80.38 cm^3
Molar Volume	= 254.7 cm^3
Parachor	= 654.6 cm^3
Index of Refraction	= 1.543
Surface Tension	= 43.5 dyne/cm
Density	= 1.190 g/cm^3
Polarizability	= 31.86 $10^{-24}cm^3$
Monoisotopic Mass	= 303.147058 Da
Nominal Mass	= 303 Da
Average Mass	= 303.3529 Da

Alkaloids of the Quinoline Group

In this sub-head we shall see the important Cinchona bases, quinine and cinchonine, together with the stricnos bases, strychnine and brucine

Quinine and Cinchonine

Cinchona or Peruvian bark, which has been used Europe since the middle of the seventeenth century in the treatment of fever is obtained from various trees of the cinchona family found mainly in Bolivia and

Peru. It contains, in addition to a tannin and quinic acid, a series of alkaloids which are closely related to one another in structure. The most important of these are quinine $C_{20}H_{24}N_2O_4$ to which the curative action of the bark is chief due, and cinchonine $C_{19}H_{-22}N_2O$.

Quinine generally crystallizes with 3 moles H_2O and in the anhydrous condition melts at 177°C; it separates from alcohol and ether in shining needles. It is present in the yellow calisaya bark to the extent of 2 to 3 per cent, has an alkaline reaction, a bitter taste and as a diacid base it forms neutral and as well as acid salts. Quinine is one of the most valuable medicines especially in the treatment of intermittent fevers such as malaria and swamp fever, and in an antidote against many infections, caused by micro organism.

Cinchonine accompanies quinine and is found in particularly large amount in grey cinchona bark (Cinchona Huanaco), in which it occurs upto 2.5 %. It crystallizes from alcohol in white prisms sublimes readily and melts at 255°C. It is used as febrifuge, but less active than quinine.

Quinine and cinchonine are similarly continued, and therefore the results obtained by the investigation of these compounds have often been supplemented one another. In many ways information gathered with regard to cinchonine has been applied without modification to quinine.

Both alkaloids were discovered in the year 1820 by Pelletier and Caventou, and their constitutional formulæ have been deduced from a large number of investigations.

As already stated, the composition of cinchonine is $C_{19}H_{22}N_2O$ and that of anhydrous quinine, as determined by Liebig is $C_{20}H_{24}N_2O_2$. In their empirical formulæ, therefore, these two bases differ in that quinine contains one atom of carbon, one atom of oxygen and two atoms of hydrogen more than cinchonine.

The molecules of cinchonine and quinine possesses two tertiary nitrogen atoms. One oxygen atom in quinine is contained in a hydroxyl and the other in a methoxyl group.

The presence of a hydroxyl group in group in quinine is indicated by several reactions. For example, quinine form a monobenzyl derivative (Schützenberger) a mono-acetyl derivative (Hesse) and a silver salt of the composition $C_{20}H_{23}AgN_2O_2$ (Straup).

The existence of a methoxyl group is shown by the formation of methyl chloride (accompanied by intramolecular rearrangement) when quinine is heated with concentrated hydrochloric acid. In this case the other primary reaction product is apoquinone, which in turn gives a diacetyl derivative and therefore contains two hydroxyl groups

$C_{19}H_{20}N_2(OH)(OCH_3)$ — Quinine

$C_{19}H_{20}N_2(OH)_2$ — Apoquinine

Proof that the oxygen atom of cinchonine is also present in hydroxyl group is supplied by reactions similar to those quoted above for quinine by the formation of acyl derivatives.

Information as to the position of the hydroxyl group and the genral structure of the cinchona alkaloids has been gained largely by the decompositions of these compounds carried out the Straup, Königs and v. Müller.

Decomposition of Quinine and Cinchonine by Fusion with Potash and by Oxidation

Fusion of these alkaloids with potash led to the conclusion that cinchonine contains a quinoline or lepidine group, and that quinine derived from 6 methoxy-lepidine.

This is in complete agreement with the results obtained by the oxidation of the alkaloids. When oxidized with a solution of chromic acid, in sulphuric acid solution, cinchonine and quinine breakup, on the one hand, into the 4-carboxylic acids of quinoline and 6-methoxy-quinoline known as cinchoninic acid and quininic acid respectively and on the other into derivatives of pyridine. Hence the molecule of the cinchona alkaloids must contain these two ring systems linked together.

The carboxylic acids of quinoline were soon identified as such, thus establishing the presence as such, thus establishing the presence of a quinoline nucleus (quinoline half) in the alkaloids.

On the other hand, the investigation of the pyridine derivatives (cincholoipon) meroquinone, cincholoiponic acid and loiponic acid proved to be exceedingly difficult and only recent been brought to a successful issue.

For a long time nothing was known of the constitution of that of the molecule giving rise to the pyridine compounds and it was briefly described by Straup as the second half of the cinchona alkaloids. From the following it will be seen that this term is for the grouping ($C_{10}H_{16}NO$), which is cinchonine is combined with the quinoline residue (C_9H_6N) – and in quinine with 6-methoxy quinoline residue ($CH_3O.C_9H_5N$).

Constitution of the "Quinoline Half" of the Quinine and Cinchonine

The key to the constitution of the "quinoline half" lies in the following facts.

1. When cinchonine is subjected to energetic oxidation, about 50% of the product consists of cinchoninic acid, which is identical with quinoline-4-carboxylic acid.

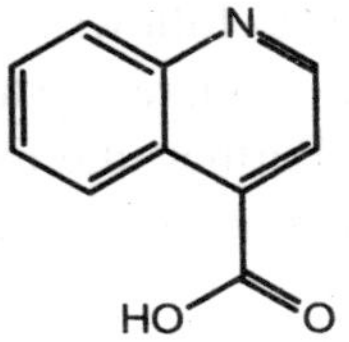

N
HO O
quinoline-4-carboxylic acid
Quinoninic acid

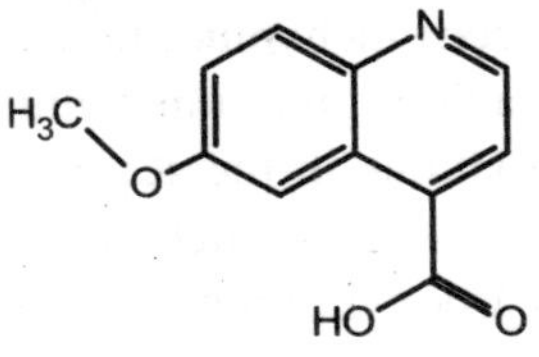

N
H_3C O
HO O
6-methoxyquinoline-4-carboxylic acid
Cinchoninic acid

From this it follows that cinchonine, containing a side chain in the 4 position. The hydroxyl group of the cinchonine is also present in the side chain because, has it been situated in the quinoline nucleus the oxidation product would have been a hydroxy-cinchoninic acid.

The composition of this side chain must be $C_{10}H_5(OH)N$ i.e. the difference between the following formula of cinchonine $C_{19}H_5(OH)N_2$ and that of the quinoline radical C_9H_6N. Hence the structure of cinchonine may be represented by formula I

[(4*E*)-8-quinolin-4-ylocta-4,7-dien-1-yl]imidoformic acid
(I)

[(4*E*)-8-(6-methoxyquinolin-4-yl)octa-4,7-dien-1-yl]imidoformic acid
(II)

2. When quinine is energetically oxidized with chromic acid it gives an acid $C_{11}H_9NO_3$ known as quininic acid. The difference between this compound and cinchoninic acid $C_{10}H_7NO_2$ is the same as that between quinine and cinchonine.

Quinic acid was shown by Straup to be the 6-methoxy-cinchoninic acid. From this it was concluded that quinine is a methoxy cinchonine, and that the methoxy group replaces the hydrogen atom corresponding to position 6 of the quinoline nucleus present in the cinchonine. Quinine was first represented by formula II.

In the above oxidations the disruption of the "second half" of quinine and the difference in the two alkaloids may be summerised in the statement that the former is a derivative of quinoline and the latter of 6-methoxy quinoline.

In formation as to the structure of the cinchona bases has also been obtained by the investigation of certain intermediate products of oxidation known as "tenines".

Straup found that the formation of cinchotenine by the oxidation of cinchonine with potassium permanganate is accomplished by the production of formic acid, according to the equation.

$$\underset{\text{Cinchonine}}{C_{19}H_{22}N_2O} + 4O \rightarrow \underset{\text{Cinchotenine}}{C_{10}H_{20}N_2O_3} + \underset{\text{Formic acid}}{HCOOH}$$

Similarly quinine, when oxidized with potassium permanganate *quitenine* and formic acid is produced.

The relationship between cinchonine and cinchotenine is established as follows; when cinchotenine is oxidized with chromic acid it gives cinchoninic acid. The tenines must therefore have been formed from cinchonine by alternations in the second half of the molecule. Further, the hydroxyl group of cinchonine is still present as such in cinchonine, although the acid properties of the latter are due to the additional presence of a carboxyl group.

Since in any case cinchonine only possesses on oxygen atom and this is contained in a hydroxyl group which is found unchanged in cinchotenine, it follows that the carboxyl must have originated in the oxidation of a group consisting solely of carbon and hydrogen.

Some information about this group is given by the following facts. Whereas cinchonine has the power of combining directly with hydrogen halide, cinchonine has not. The formation of the carboxyl group with unsaturated properties. During the destruction of this group only one carboxyl group is formed, and a carbon atom is detached as formic acid. Further, cinchotenine is easily transformed into cincholoiponic acid, which must be regarded as a side chain. All these facts can only be satisfactorily explained by assuming that cinchonine contains a vinyl group. This group unites with hydrogen halides, and on oxidation is ruptured at the double bond with the production of a carboxyl group and formic acid.

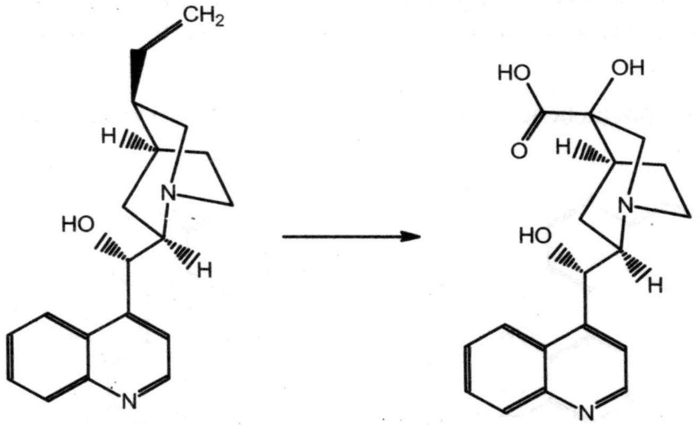

cinchonine → cinchotenine

The conclusions just arrived at for cinchonine can also be applied directly to quitenine. The property processed by quinine and cinchonine of uniting with two atoms of bromine, or a molecule of halogen, is in complete agreement with the above constitution.

Constitution of the "Second Half" of Quinine and Cinchonine

Valuable information concerning the constitution of the "second half" of the cinchona bases has been obtained by Königs who examined the hydrolysis products of cinchonine and quinine.

Cinchene, $C_{19}H_{20}N_2$ is the anhydrous compound of cinchonine $C_{19}H_{22}ON_2$ **and quinine,** $C_{19}H_{19}(OCH_3)N_2$ is the anhydro-compound of **quinine** $C_{19}H_{22}(OCH_3)N_2O$.

In many respects, these compounds are much more reactive than the parent alkaloids from which they are obtained by successive treatment with phosphorous pentachloride and alcoholic potash.

Hydrolytic Decomposition of Cinchene and Quinene

According to experimental conditions cinchene and quinine may take up the elements of water and decompose in two distinct ways.

1. **When the anhydrous** – bases are boiled for a long time with concentrated hydrobromic acid, they take up one molecule of water and at the same time lose of a molecule of ammonia.

$$C_{19}H_{20}N_2 + H_2O \rightarrow C_{19}H_{19}NO + NH_3$$

$$C_{19}H_{19}(OCH_3)N_2 + H_2O + HBr \rightarrow C_{19}H_{19}NO_2 + NH_3 + CH_3Br$$

In this manner they yield derivative of 4-phenyl-quinoline, which were described by Königs and Comstock as apocinchene and apoquinone respectively.

The information gained from the investigation of apocinchene can be directly applied to the constitution of apoquinone, since the latter is easily converted into the former. Thus when apoquinone is heated at 250°C with a mixture of ammonium chloride and the soluble compound of ammonia with zinc chloride, only the hydroxyl group present in the quinoline residue is replaced by an amino group. The amino apoquinone so obtained can be then be converted through the azo compound into apocinchene

Königs showed that apocinchene were derivative of 4-O hydroxyphenyl-quinoline. They have the following constitutions:

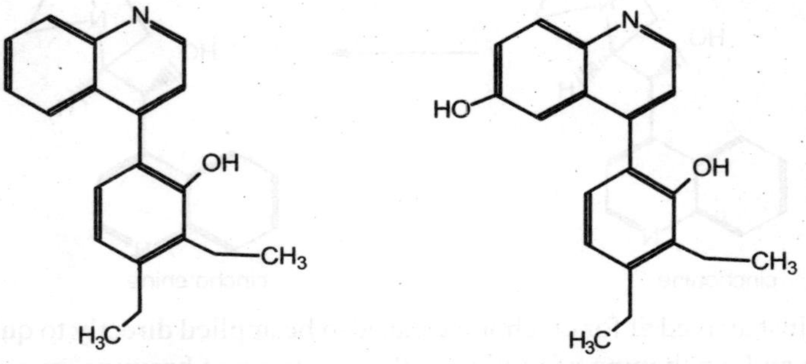

Apochinchene Apoquinene

The conversion of apoquinone into apocinchene outlined above leads to the conclusion that apoquinone leads to the conclusion that apoquinone is hydroxy-apocinchene, and when taken in conjunction with the facts given before proves that the additional hydroxyl group must occupy position 6 in the quinoline nucleus. The complete analogy in the behaviour of quinene and cinchene points to the analogous structure of the two anhydro bases. Hence it may be derived from cinchene by replacement of the hydrogen atom in position 6 in the quinoline nucleus by a methoxyl group.

2. When cinchene is heated with 20 percent aqueous phosphoric acid at 170 to 180°C; it yields lepidine and a compound of the composition $C_9H_{15}NO_2$ described by Königs as *meroquinene.* The hydrolysis of cinchene under these conditions therefore takes place according to the equation.

$$\underset{\text{Cinchene}}{C_{19}H_{20}N_2} + 2H_2O \rightarrow C_{10}H_9N + \underset{\text{Meroquinene}}{C_9N_{15}NO_2}$$

Constitution of Meroquinine, Cincholoiponic Acid and Loiponic Acid

Königs found that meroquinene could be obtained by the direct oxidation of cinchonine with chromic acid, as well as by the hydrolysis of cinchene and quinene. On further treatment with an ice-cold aqueous solution of potassium permanganate and sulphuric acid, meroquinene yields *cincholoiponic acid.* This compound was also obtained by Straup and by the direct oxidation of cinchonine and quinine

$$\underset{\text{Meroquinone}}{C_9H_{15}NO_2} + 4O \rightarrow \underset{\text{Cincholoiponic acid}}{C_8H_{13}NO_4} + HCOOH$$

From cincholoiponic acid, by careful oxidation with potassium permanganate, Straup isolated very small quantities of loiponic acid. Hence meroquinone, cincholoiponic acid, loiponic acid represent successive stages in the oxidation of the "second half" of cinchonine. Königs formulates the compounds as follows:

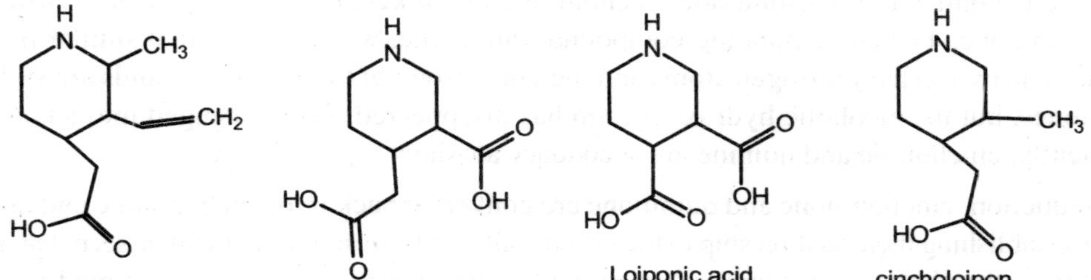

Meroquinene cincholoiponic acid Loiponic acid cincholoipon

On reduction with zinc dust and hydriodic acid, meroquinene is converted into *cincholoipon* the formula for which is also given above. The presence of the carboxyl group in these products has been proved by the preparation of esters, and that of the imino group by the preparation and properties of nitrosamines, and a acetyl and N-alkyl detrivatives. The position of the carbonyl groups in meroquinene and cincholoipon where finally established as a result of the synthesis of ethyl quinuclidine, which is going to be discussed.

The assumption that these compounds contained a pyride nucleus was based primarily on the formation of 4-methyl-2-ethyl pyridine. When meroquinene is heated with mercuric chloride in hydrochloric acid and also on the conversion of cincholoiponic acid into 4-methyl-pyridine by means of concentrated sulphuric acid.

The most convincing proof of the presence of a pyridine nucleus in loiponic acid is due of Königs, when this compound is heated with potassium hydroxide it is transformed into an isomeric acid, which is identical with synthetic *hexahydro-cinchomeronic acid (piperidine – 3:4 dicarboxylic acid)* Loiponic acid is therefore a labile form of hexahydrocinchomeronic acid, which passes into the stable form on being heated with alkali.

The formula of cincholoiponic acid was eventually confirmed by Wöhl who succeeded in synthesizing both of the theoretically possible racemic compounds from β-chloro-propionacetal. These racemic compounds where resolved into the four optically active forms, one of which proved to be identical in all respects with the cincholoiponic acid obtained quinine by Straup.

The additional knowledge of the constitution of cinchonine and quinine gained by the study of meroquinene, cincholoipon cincholoiponic acid and loiponic acid may therefore be summerised as follows:

Of the ten carbon atoms present in the "second half" of the cinchona bases, five are contained in a piperidine nucleus, two in a vinyl group. The points at which the vinyl and methyl groups are attached to the piperidine nucleus, two in a vinyl and methyl groups are attached to the piperidine nucleus have been determined.

The hydrolysis of cinchene and quinene to give meroquinene on the one hand and lepidine or methoxy lepidine on the other, indicates that piperidine and quinoline nuclei are united by another carbon atom.

By means of moderate oxidation with chromic acid in acetic acid or sulphuric acid solution, Rabe converted cinchonine and its isomeride cinchonidine into a ketone, which by analogy with tropinone was cinchoninone. The corresponding compound, quininone, was obtained in a similar manner from quinidine. The two tertiary nitrogen atoms and the vinyl group of the parent alkaloids are still present in these ketones, but the alcoholic hydroxyl group has disappeared, being changed into a ketonic group. Consequently, cinchonine and quinine are secondary alcohols.

On reduction, cinchoninone and quininone are converted back into cinchoninone and quinine, thus definitely establishing their relation ship to the original alkaloids. In addition, it will be seen that information obtained from the chemical decomposition of the ketones may be applied without modification to the cinchona alkaloids themselves.

Constitution of Quinine and Cinchonine

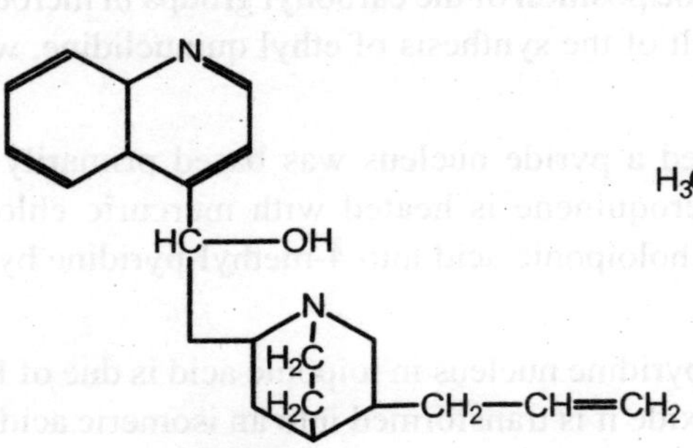

cinchonine

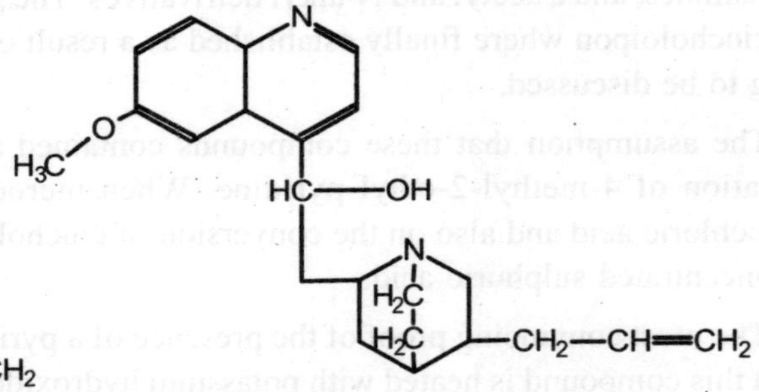

Quinine

Since each of the above mentioned formula contains three asymmetric carbon atoms, each structure represents a possibility of eight optically active and four racemic stereoisomerides. Among these are cinchonidine which is a stereoisomeric with cinchonine, and conchinine, stereoisomerides of quinine.

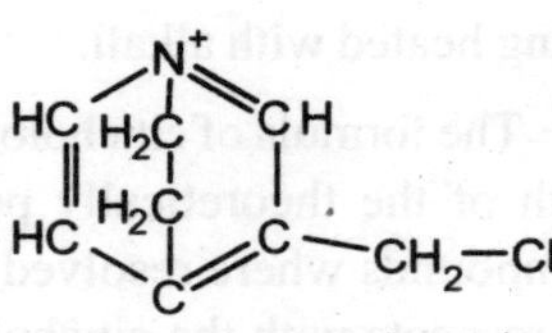

The possibility of the existence of carbon bridge between the nitrogen atom and the 4 carbon atom of the annexed structure. The parent compound quinuclidine has also been synthesized.

β - Ethyl quinuclidine

Synthesis of Cinchona Alkaloids

When cinchonine and quinine are boiled in acetic acid for a long time, they undergo intramolecular change and yield cinchotoxine and quinotoxine respectively. These toxins are formed from the alkaloids by the conversion of %COOH group into %CO, accompanied by the rupture of the quinuclidine nucleus. They are ketones as well as secondary bases. The reverse changes from the toxins into the alkaloids has not been accomplished directly as yet, but has been effected indirectly through the following stages. The hydrogen atom of the imino group can be replaced by bromine to give bromo-imines which by loss of a molecule of hydrogen bromine can be converted into the compound quinone and cinchoninone.

These reactions lead to the regeneration of the quinuclidine nucleus alkaloids. Finally, the ketones can be reduced to the alkaloids themselves.

The partial synthesis of quinine from quinotoxine or quinicine by this method proceeds as follows:

Quinicine (I) is converted into N-bromoquinicine (II) by the action of sodium Hypobromite. Quininone (III) is obtained from the bromo-imine by treatment with alkali, and is then reduced to quinine (IV) by reduction in alcoholic solution using aluminium powder in the presence of sodium ethylate.

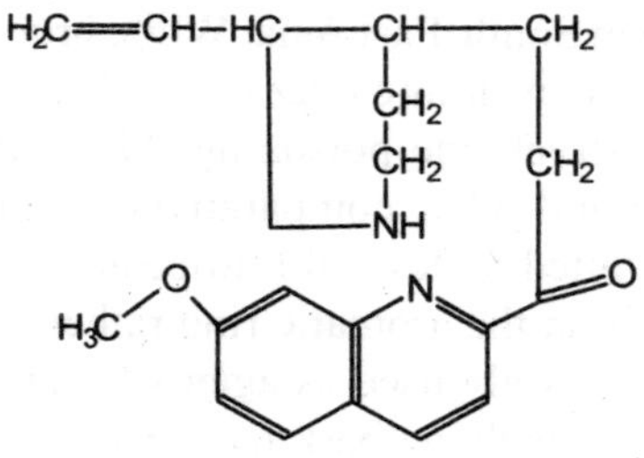

(I) 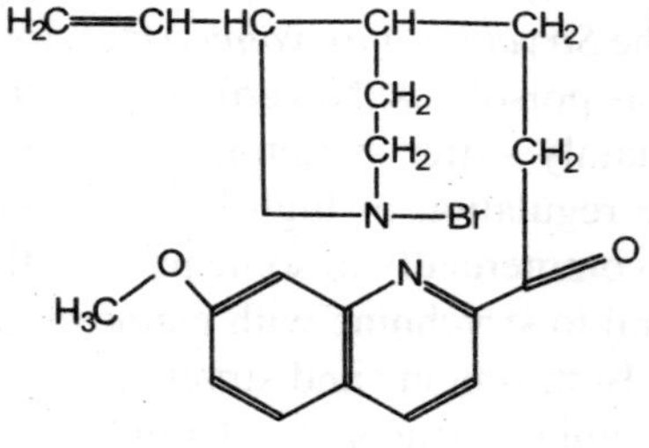 (II) 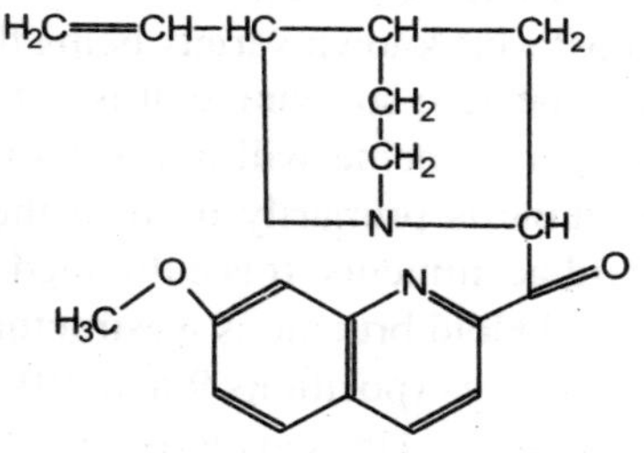 (III)

The above investigations established the position of the carbonyl group in the toxines, and revealed methods of opening up the *quilidine nucleus* of the alkaloids, as well as of converting the toxines so obtained back into the parent compounds. This measure of success paved the way for a number of attempts to synthesize cinchonine and quinine, and bases closely related to them. In every case the starting material was either cinchoninic acid, quininic acid, or an ester or nitrile of these acids. By making use of a convenient method of preparing cyano-quinilines, Kaufmann succeded in synthesizing various acids and 4-quinonyl ketones and later obtained bases of the types V which were closely related to the cinchona bases.

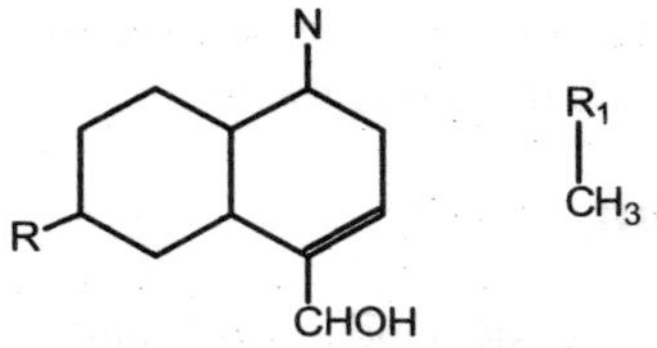

R = HOR (alkyl group)

R_1 = alkyl group

The synthesis of cinchona alkaloids from derivatives of piperidine and quiniline has been completed by Rabe, using cyano-quinolines and cinchonine ester as the starting materials. From 4-methyl-pyridine (obtained from coal tar) Rabe synthesized cincho – and quinio toxines which however, contained no vinyl group.

In conclusuin, it may be mentioned that hydro-derivatives such as dihydro-cinchona bark. They may be prepared from the last-named alkaloids by hydrogenation.

The Strichnos Alkaloids

There are three alkaloids in this series, namely strychnine, brucine.

Strychnine is a very toxic (LD_{50} = 10 mg approx.), colorless crystalline alkaloid used as a pesticide, particularly for killing small vertebrates such as rodents. Strychnine causes muscular convulsions and eventually death through asphyxia or sheer exhaustion. The most common source is from the seeds of the *Strychnos nux vomica* tree. Strychnine is one of the most bitter substances known. Its taste is detectable in concentrations as low as 1 ppm.Strychnine acts as a blocker or antagonist at the inhibitory or strychnine-sensitive glycine receptor (GlyR), a ligand-gated chloride channel in the spinal cord and the brain.

Brucine is a bitter alkaloid closely related to strychnine. It can be found in some plant species, the most well-known variety being the *Strychnos nux-vomica* tree, found in South-East Asia. While brucine is related to strychnine, it is not as poisonous. Nevertheless, a human consuming over two milligrams of pure brucine will almost certainly suffer symptoms resembling strychnine poisoning. Medically, brucine is primarily used in the regulation of high blood pressure and other comparatively benign cardiac ailments. It is cultivated commercially in some parts of the United States and European Union. he alkaloid brucine is isostructural to strychnine with methoxy groups at the aromatic ring rather than hydrogens (positions 9 and 10). Both brucine and strychnine are commonly used as agents for chiral resolution. The separation of racemic mixtures by alkaloids from the cinchona bark has been known since 1853 when its use as such was reported by Pasteur. The ability of brucine, and to a lesser extent strychnine, to function as resolving agents for amino acids was reported by Fisher in 1899. Brucine and strychnine are basic and thus have a tendency to crystallise with acids. The acid-base reaction leaves the brucine protonated at the N(2) position. The formation of diastereomeric salts has been reported for thousands of organic compounds. The packing of brucine in corrugated (waving) layers was an essential aspect in the co-crystallisation of brucine, whereas strychnine was shown to crystallise predominantly in bilayers.

Test of brucine

Brucine is detected by mixing it with conc. sulphuric acid and then a drop of nitric acid is added. Which turns into a deep orange-red colour, this changes to violet on the addition of stannous chloride and water. This is a test of nitric acid and nitrate even at minute traces

Alakloids of the Isoquinoline Group

This group includes the five opium alkaloids, papaverine, laudanosine laudanine, narcotine and narceine. All these are known to be related to isoquinoline, the first three being comparatively simple derivatives.

When it is remembered that the alkaloids hydrastine and berberine found in the root of *hydrastis Canadensis*, are also derived from iso-quinoline, the importance of the latter as a parent compound of alkaloid.

PAPAVERINE

Introduction

Papaverine is an opium alkaloid used primarily in the treatment of visceral spasm, vasospasm (especially those involving the heart and the brain), and occasionally in the treatment of erectile dysfunction. While it is found in the opium poppy, papaverine differs in both structure and pharmacological action from the other opium alkaloids (opiates).

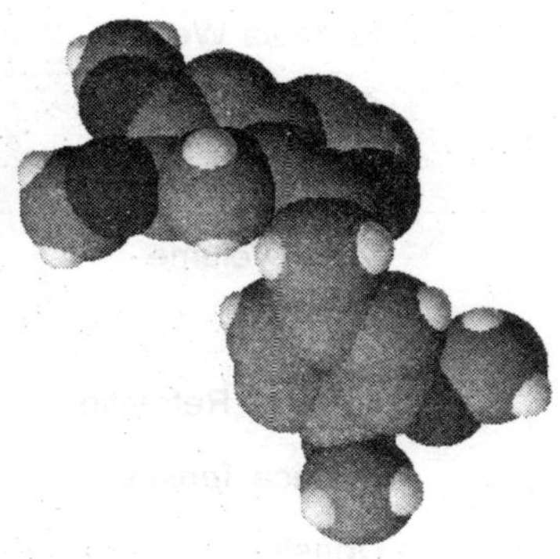

Synthesis

The synthesis of papaverine was accomplished by A Pictet and Gams by the reactions summerised below:

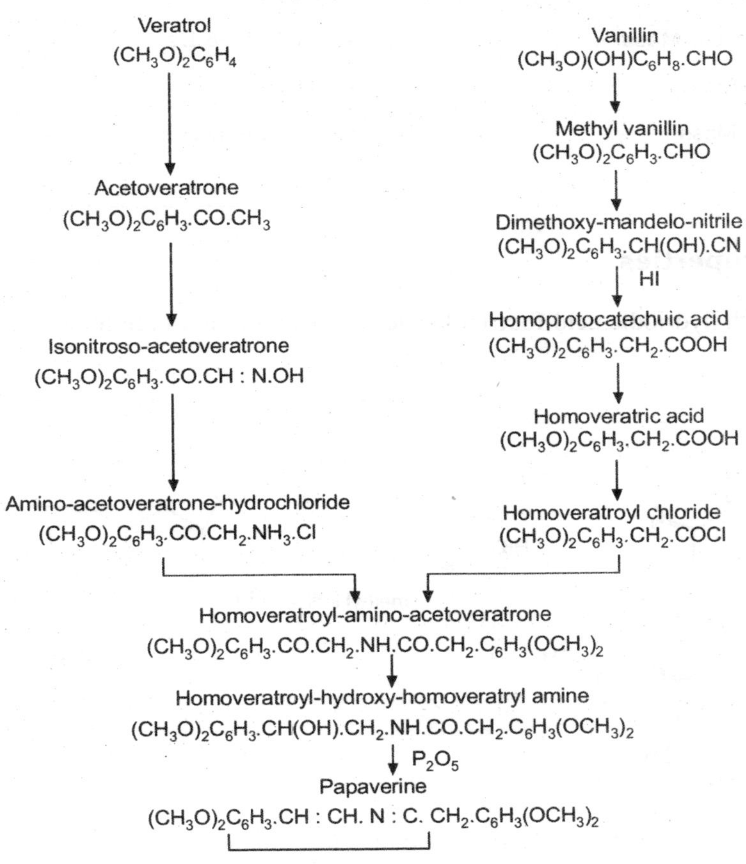

Physical Properties

Molecular Formula	$= C_{20}H_{23}NO_4$
Formula Weight	= 341.40092
Composition	= C(70.36%) H(6.79%) N(4.10%) O(18.75%)
Molar Refractivity	= 95.48 cm^3
Molar Volume	= 287.8 cm^3
Parachor	= 760.1 cm^3
Index of Refraction	= 1.577
Surface Tension	= 48.6 dyne/cm
Density	= 1.18 g/cm^3
Polarizability	= 37.85 $10^{-24}cm^3$
Monoisotopic Mass	= 341.162708 Da
Nominal Mass	= 341 Da
Average Mass	= 341.4009 Da
Log P	= 3.02

Chemical Properties

When heated with hydriodic acid four molecules of methyl iodides are liberated and papaveroline is produced.

H_3C H_3C-O N CH_3 CH_3

Papaverine

$\xrightarrow[\Delta]{HIO_3}$ H_3C—I

methyl iodide

HO HO CH_2 OH HO

papaveroline

The reaction proves that papaverine constitutes four methoxy groups.

Decomposition of Papaverine is decomposed into dimethoxy-isoquinoline and a compound which does not contain nitrogen. The latter has been shown to be *dimethyl homocatecol* as it yields protocatechuric acid on

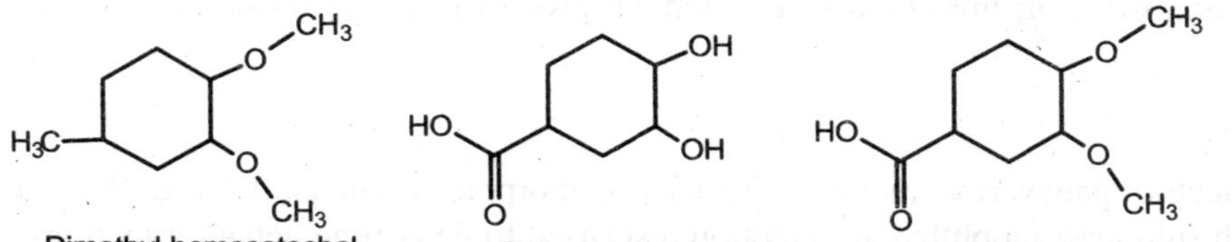

Dimethyl homocotechol   Protocatechuric acid   3,4-dimethoxycyclohexanecarboxylic acid

more energetic treatment with potash. In addition, an appreciable quantity of veratric acid is always produced during the oxidation of the alkaloid. It will be observed that the side chains occupy the same position in all these compounds.

Constitution of Papaverine

Papaverine can therefore be locked upon as a combination of dimethoxy-isoquinoline and dimethyl homocatechol

$$C_{11}H_{11}NO_2 \quad + \quad C_9H_{12}O_2 \quad \rightarrow \quad C_{20}H_{21}NO_4 \quad + \quad H_2O$$

The manner in which these two components are joined together was determined by Goldschmidt in the following way:

Papaverine contains four methoxy groups, two of which are contained in each of which are contained in each of the above disruption products. Since the components cannot be united through the methoxy groups, they must be joined through carbon atom of the benzene nucleus, or of the methyl group of dimethyl homocatechol. The latter supposition is supported by the whole behaviour of the papaverine, especially the case with which the component parts can be separated. Hence, the alkaloid is a substituted *phenyl-isoquinoline methane.*

Finally, it was necessary to determine which carbon atom of the isoquinoline ring takes part in the union. This point is decided by the fact that when papaverine is oxidized with potassium permanganate, *α-carboncinchomeronic acid (2:3::4 – pyridine) tricarboxylic acid* is formed.

Uses

Papaverine is approved to treat spasms of the gastrointestinal tract, bile ducts and ureter and for use as a cerebral and coronary vasodilator in subarachnoid hemorrhage (combined with balloon angioplasty) and coronary artery bypass surgery. Papaverine may also be used as a smooth muscle relaxant in microsurgery where it is applied directly to blood vessels.

The *in vivo* mechanism of action is not entirely clear, but an inhibition of the enzyme phosphodiesterase causing elevation of cyclic AMP levels is significant. It may also alter mitochondrial respiration.

It is also commonly used in cryopreservation of blood vessels along with other glycosaminoglycans and protein suspensions. Functions as a vasodilator during cryopreservation when used in conjunction with verapamil, phentolamine, nifedipine, tolazolines, or nitroprusside.

Papaverine is also being investigated as a topical growth factor in tissue expansion with some success.

Side Effects

Frequent side effects of papaverine treatment include polymorphic ventricular tachycardia, constipation, interference with sulphobromophthalein retention test (used to determine hepatic function), increased transaminase levels, increased alkaline phosphatase levels, somnolence, and vertigo.

Rare side effects include flushing of the face, hyperhidrosis (excessive sweating), cutaneous eruption, arterial hypotension, tachycardia, lack of appetite, jaundice, eosinophilia, thrombopenia, mixed hepatitis, headache, allergic reaction, chronic active hepatitis, and paradoxical aggravation of cerebral vasospasm.

NARCOTINE

Introduction

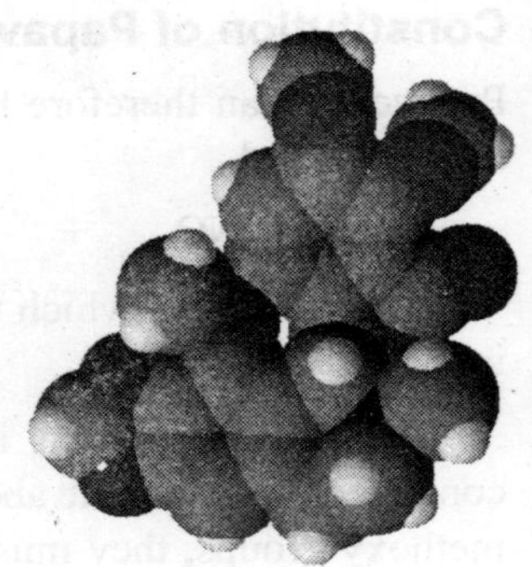

This alkaloid is freely present in the opium, in morphine and codeine from opium by extraction with water, narcotine is obtained by treating the residue with warm ether. It crystallizes and has a melting point of 176°C and is insoluble in cold water and alkali

Physical Properties

Molecular Formula	= $C_{22}H_{23}NO_7$
Formula Weight	= 413.42052
Composition	= C(63.91%) H(5.61%) N(3.39%) O(27.09%)
Melting Point	= 176^0C
Molar Refractivity	= 106.49 cm^3
Molar Volume	= 310.2 cm^3
Parachor	= 826.5 cm^3
Index of Refraction	= 1.602
Surface Tension	= 50.4 dyne/cm
Density	= 1.332 g/cm^3
Polarizability	= 42.21 $10^{-24}cm^3$
Monoisotopic Mass	= 413.147452 Da
Nominal Mass	= 413 Da
Average Mass	= 413.4205 Da

Chemical Properties

Narcotine is a *tertiary base* and does not react with acetic acid anhydride hence the molecule contains no free hydroxyl group.

The presence of *three methoxyl group* is shown by heating the portions of methyl chloride with nonarcotine $C_{19}H_{17}NO_7$ or $C_{19}H_{14}NO_4(OH)_3$.

Narcotine is decomposed by potassium hydroxide at 220°C with liberation of methyl amine, dimethylamine as trimethylamine; the nitrogen atom is therefore attached to a methyl group.

Our knowledge of the constitution of narcotine is largely due to the work of Koser.

Decomposition of Narcotine into Nitrogenous and Nitrogen Free Components

The decomposition of narcotine under the influence of reagents such as water at about 140°C, dilute acids and alkalies, has thrown valuable light on its composition. Under this treatment it yields nitrogen-free compound opianic acid and a base hydrocotarnine.

$$\underset{}{C_{22}H_{23}NO_7} + H_2O \rightarrow \underset{\text{Opianic acid}}{C_{10}H_{10}O_5} + \underset{\text{base hydrocotarnine}}{C_{12}H_{15}NO_3}$$

With oxidizing agents like nitric acid, platinic chloride, ferric chloride or lead peroxide, narcotine decomposes in a similar manner to give opianic acid and cotarnine.

$$C_{22}H_{23}NO_7 + O + H_2O \rightarrow \underset{\text{Opianic acid}}{C_{10}H_{10}O_5} + \underset{\text{Cotarnine}}{C_{12}H_{15}NO_4}$$

Reducing agents such as zinc and hydrochloric acid, and sodium amalgam, convert narcotine into *meconine* which is a reduction product of opianic acid, and hydro-cotarnine.

$$C_{22}H_{23}NO_7 + H_2 \rightarrow \underset{\text{Meconine}}{C_{10}H_{10}O_4} + \underset{\text{Hydrocotarnine}}{C_{12}H_{15}NO_3}$$

From these reactions it is evident that the molecule of its evident that the molecule of narcotine consists essentially of two parts; a basic components corresponding to hydrocotarnine, and a nitrogen-free substance corresponding to opianic acid.

Once the structure of these individual parts has been determined, it was possible to build up the formula of narcotine itself.

Opianic acid was investigated Beckett and Wright, also by Wegscheider, proved to be a carboxylic acid, and derived from dimethyl protocatechiic aldehyde.

The molecule of narcotine consists esentially of two parts; a basic components corresponding to hydrocotarnine, and a nitrogen-free substance corresponding to opianic acid.

From a study of the p products obtained by oxidation, and by interaction with methyl iodide, cotarnine has been assigned formula II (Roser). From an alternative formula see (IV).

(I) (II)

(III) (IV)

MECONINE

Introdcution

Meconine (III) has been shown to be the lactone of the alcohol corresponding to opianic acid.

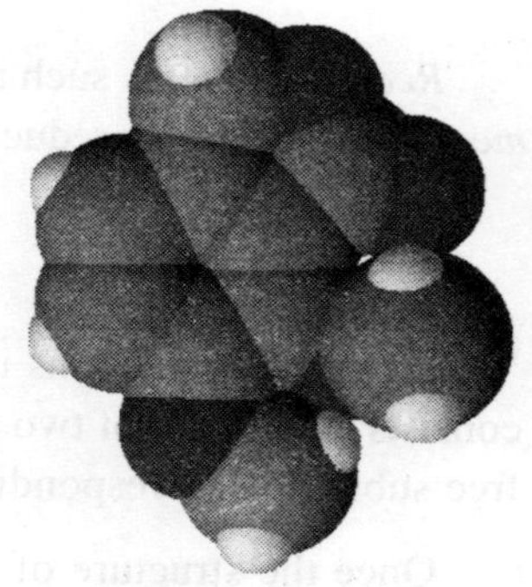

The conversion of cotarnine into hydrocotarnine, by means of reducing agents, takes place by reduction of the aldehyde group –CHO to CH_2OH, followed by the formation of an isoquinoline ring by hydro – cotarnine is therefore assigned formula (IV).

The constitution of the decomposition products opianic acid and meconine on the other hand, and cotarnine and hydro –cotarnine on the other, is therefore clear. The only point still to be decided is how the two component parts are linked together to form the alkaloid molecule.

The narcotine, $C_{22}H_{23}NO_7$, the hydro-cotarnine group cannot be united with opianic acid or meconine through one of the seven oxygen atoms, since five of these are already joined to alkyl groups (three to methyl and two to methylene) and the other two are both present in a lactone ring. Further, the valances of the nitrogen atom are fully satisfied by the demands of the isoquinoline ring and the methyl group. The two components must therefore be connected through carbon atoms. There is no doubt that it is these carbon atoms which take up oxygen during the oxidation of the alkaloid and appear as aldehyde group is present in narcotine itself. For these reasons narcotine is given the constitution stated previously.

Synthesis of Meconine

Guaiacol carboxylic acid, on methylation was converted into the methyl ester of 2:3 dimethoxy benzoic acid, I methyl phthalide, II, which was then hydrolysed to the corresponding phthalide carboxylic acid. The acid, on being heated, decomposed into carbon dioxide and meconine.

2:3 dimethoxy benzoic acid + trichloroacetaldehyde → 5:6 dimethoxy trichloro methyl phthalide + methanol

Which was then hydrolysed to the corresponding phthalide carboxylic acid. The acid, on being heated, decomposed into carbon dioxide and meconine.

Physical Properties

Molecular Formula	$= C_{10}H_{10}O_4$
Formula Weight	= 194.184
Composition	= C(61.85%) H(5.19%) O(32.96%)
Molar Refractivity	= 48.90 cm^3
Molar Volume	= 154.0 cm^3
Parachor	= 395.1 cm^3
Index of Refraction	= 1.547
Surface Tension	= 43.2 dyne/cm
Density	= 1.260 g/cm^3
Polarizability	= 19.38 $10^{-24}cm^3$
Monoisotopic Mass	= 194.057909 Da
Nominal Mass	= 194 Da
Average Mass	= 194.184 Da

Uses

It is generally used in the medicine especially in expectorants and mild sedatives.

Synthesis of Narcotine

Perkin and Robinson shown that when meconine and cotarnine are boiled in alcoholic solution with the racemic alkaloid *gnoscapine*. By resolving this into its active components *d* and *l*-narcotine were obtained, the *l*-variety being identical with the natural alkaloid.

Synthesis of Cotarine

This base was synthesized by Salway and later by Decker and Becker, from myristicin, a constituent of oil of parsley and oil of nutmey. An intermediate product in the latter synthesis was formyl-homomyristicyl amine.

Myristican → Formyl homomyristicyl → Cotarnine hydrochloride → cotarnine

Spectroscopic investigation by Dobbie, Lauder and Tinkler have shown that cotarnine salts correspond to the ammonium structure (III) but that the free base may exist as the carbonium forms (in aqua or alcoholic solution)

Narceine is obtained by the action of alkali on narcotine methiodide. It is present in opium in small quantities (about 1%) and is a white crystalline compound of melting point 171°C.

Narcine

It has the proceeding structure, and is formed from narcotine by the rupture of the pyridine and lactone ring.

HYDRASTINE

Introduction

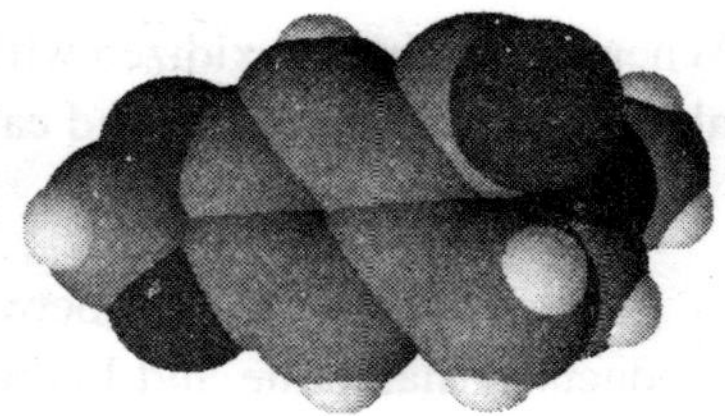

Hydrastine is a natural alkaloid which was discovered in 1851 by Alfred P. Durand. By hydrolysis of Hydrastine one receives Hydrastinine, which was patented by Bayer as a haemostatic drug during 1910s.

Synthesis

Hydrastine, available from the roots of the Hydrastis candensis, a plant belongs to the Ranunculaccæ and indigenous to North America. It crystallizes in prisms. The extract of Hydrastis candensis is used therapeutically in cases of uterine hemorrhage and gastritis in homeopathy.

Hydrastine is a derivative of Piperonal consisting a basic side chain in the ortho position to the aldehyde group. When it is reduced with zinc and hydrochloric acid, ring formation takes place with loss of oxygen and production of hydrogen hydrastine. This compound is an isoquinoline derivative and formed an intermediate product in first synthesis of hydrastis to be effected.

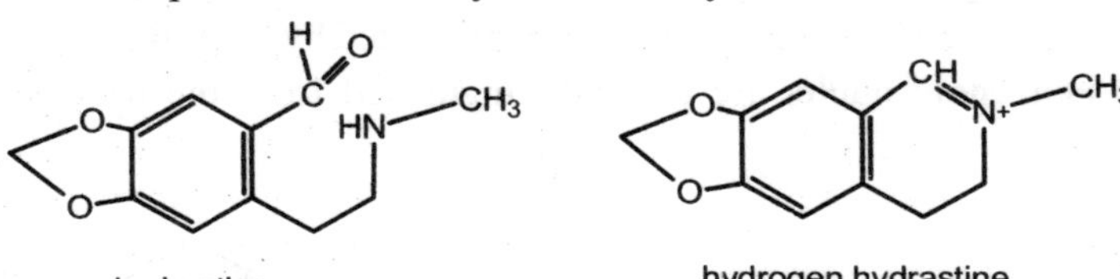

Physical Properties

Molecular Formula	= $C_{11}H_{13}NO_3$
Formula Weight	= 207.22582
Composition	= C(63.76%) H(6.32%) N(6.76%) O(23.16%)
Melting Point	= 135°C
Molar Refractivity	= 56.94 cm^3
Molar Volume	= 170.4 cm^3
Parachor	= 452.0 cm^3
Index of Refraction	= 1.582
Surface Tension	= 49.5 dyne/cm
Density	= 1.216 g/cm^3
Polarizability	= 22.57 $10^{-24}cm^3$
Monoisotopic	= 207 Da
Average Mass	= 207.2258 Da

Chemical Properties

When hydrastine is oxidized with potassium permanganate in acid solution then opianic acid produced along with a basic compound called hydrstinine.

$$C_{21}H_{21}NO_6 + H_2O + O \rightarrow C_{10}H_{10}O_5 + C_{11}H_{10}O_5 + C_{11}H_{13}NO_3$$

The difference of CH_2O between the molecule of cotarnine and hydrastinine, the basic decomposition products of narcotine and hydrastine respectively, indicates that cotarnine is methoxy-hydrastinine. Narcotine must therefore be a methoxy hydrastine with the methoxyl group confirmed by the determination of the methoxyl group by Zeisel's Method. Hydrastinine is of the greatest importance in connection with the constitution of the hydrastine. Its structure has been ascertained both by degradation and synthesis.

Alakloids of the Phenanthrene Group

A number of the above-mentioned alkaloids, including *glaucine*, *bulbocapine* and *corytuberine* are derivatives of aporphine, a base containing a condensed phenathrene – pyridine structure. Glaucine and aporphine have been synthesized by Gadmar; they are represented by the following abbreviated formula, in which the normal benzenoid nuclei are to be distinguished from the hydrogenated rings (Me = methyl)

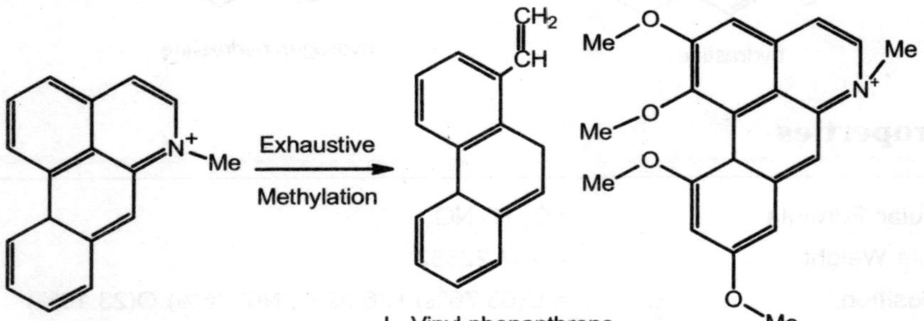

Exhaustive Methylation

Aporphine

I - Vinyl phenanthrene

Glaucine

From their general properties and behaviour on oxidation the last two compounds has been assigned.

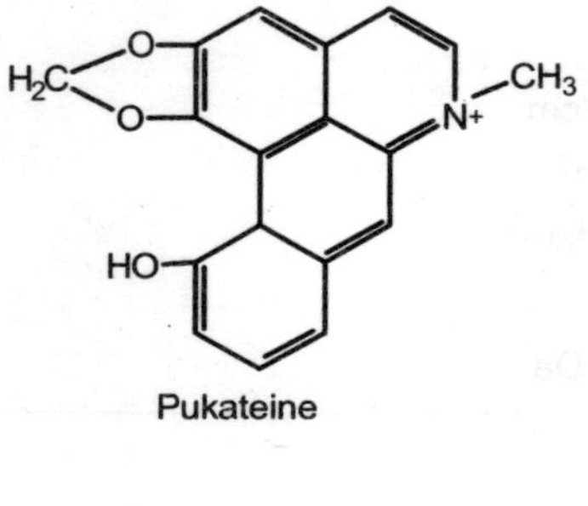

Pukateine

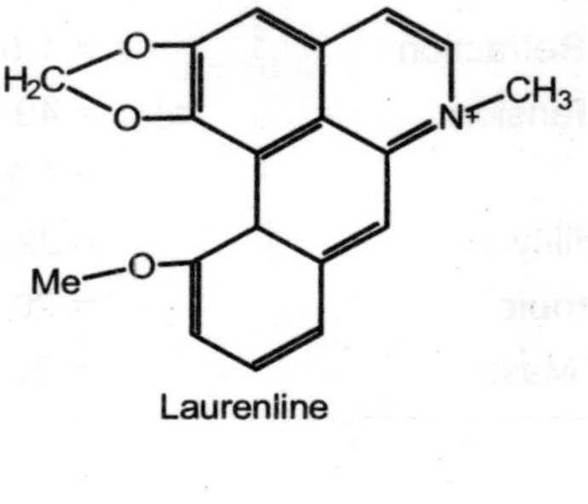

Laurenline

The above constitutions in the respect to physiological action represent morphine. When subjected to exhaustive methylation aporphine yields I-vinyl phenanthrene. The methoxyl derivatives under similar treatment are converted into the corresponding I-methyl methoxy-phenanthrene.

MORPHINE

Morphine was first isolated in 1804 in Paderborn, Germany by the German pharmacist Friedrich Wilhelm Adam Sertürner, who named it "morphium" after Morpheus, the Greek god of dreams. But it was not until the development of the hypodermic needle in 1853 that its use spread. It was used for pain relief, and as a "cure" for opium and alcohol addiction. Later it was found out that morphine was even more addictive than either alcohol or opium, and its extensive use during the American Civil War allegedly resulted in over 400,000 sufferers from the "soldier's disease" of morphine addiction. This statement has been subjected to controversy, as there have been suggestions that such a disease was in fact a hoax and soldier's disease never occurred after the Civil War.

N—CH_3

HO O OH

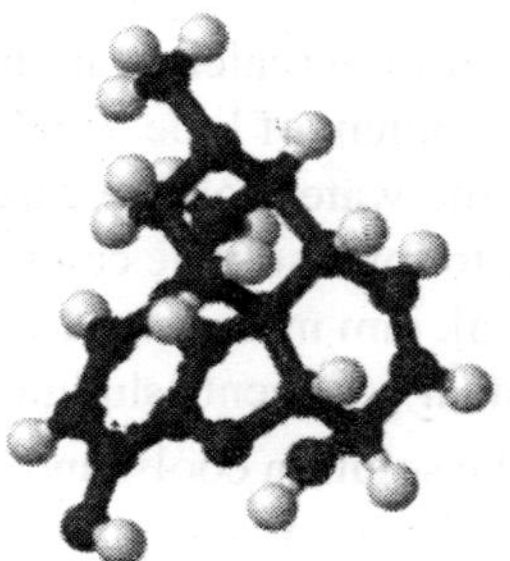

Diacetylmorphine (heroin) was derived from morphine in 1874 and brought to market by Bayer in 1898. Heroin is approximately 1.5-2 times more potent than morphine on a mg for mg basis. Using a variety of subjective and objective measures, the relative potency of heroin to morphine administered intravenously to post-addicts found 1.80 mg of morphine sulfate equals to 1 mg of diamorphine hydrochloride (heroin). The pharmacology of heroin and morphine is identical except the two acetyl groups increase the lipid solubility of the heroin molecule, and thus the molecule enters the brain a bit more rapidly. The additional groups are then detached, yielding morphine, which is what causes the subjective effects of heroin. Therefore, the effects of morphine and heroin are identical except that heroin is slightly more potent and acts slightly faster. Morphine, along with heroin and cocaine were outlawed and their possession without a prescription was criminalized in the U.S. by the Harrison Narcotics Tax Act of 1914.

In 1952, Dr. Marshall D. Gates, Jr. was the first person to chemically synthesize morphine at the University of Rochester. This breakthrough is well renowned in the field of organic chemistry.

This entire process is only intended for the study, research and experimental purpose not for any bad or dirty use. Students should keep this in mind that after reading this book they are going to be chemists and not a drug maffia.

Morphine is routinely carried by soldiers on operations in an autoinjector.

Morphine was the most commonly abused narcotic analgesic in the world up until heroin was synthesized and came into use. Even today, morphine is the most sought after prescription narcotic by heroin addicts when heroin is scarce.

Synthesis of Morphine from Opium Popy

The following is a step-by-step description of morphine extraction:

An empty 55-gallon oil drum is placed on bricks about a foot above the ground and a fire is built under the drum. Thirty gallons of water are added to the drum and brought to a boil. Ten to fifteen kilograms of raw opium are added to the boiling water.

1. With stirring, the raw opium eventually dissolves in the boiling water, while soil, leaves, twigs, and other non-soluble materials float in the solution. Most of these materials are scooped out of the clear brown 'liquid opium' solution.
2. Slaked lime (calcium hydroxide), or more often a readily available chemical fertilizer with a high content of lime, is added to the solution. The lime converts the water insoluble morphine into the water soluble calcium morphenate. The other opium alkaloids do not react with the lime to form soluble calcium salts. Codeine is slightly water soluble and gets carried over with the calcium morphenate in the liquid. For the most part, the other alkaloids become part of the residual sediment 'sludge' that comes to rest on the bottom of the oil drum.
3. As the solution cools, and after the insolubles precipitate out, the morphine solution is scooped from the drum and poured through a filter of some kind. Burlap rice sacks are often used as filters. They are later squeezed in a press to remove most of the solution from the wet sacks. The solution is then poured into large cooking pots and re-heated, but not boiled.
4. Ammonium chloride is added to the heated calcium morphenate solution to adjust the alkalinity to a pH of 8 to 9, and the solution is then allowed to cool. Within one or two hours, the morphine base and the unextracted codeine base precipitate out of the solution and settle to the bottom of the cooking pot.
5. The solution is then poured off through cloth filters. Any solid morphine base chunks in the solution will remain on the cloth. The morphine base is removed from both the cooking pot and from the filter cloths, wrapped and squeezed in cloth, and then dried in the sun. When dry, the crude morphine base is a coffee-coloured powder.
6. This 'crude' morphine base, commonly known by the Chinese term p'i-tzu throughout Southeast Asia, may be further purified by dissolving it in hydrochloric acid, adding activated charcoal, re-heating and re-filtering. The solution is filtered several more times, and the morphine (morphine hydrochloride) is then dried in the sun.
7. Morphine hydrochloride (still tainted with codeine hydrochloride) is usually formed into small brick-sized blocks in a press and wrapped in paper or cloth. The most common block size is 2 inches by 4 inches by 5 inches weighing about 1.3 kilograms (3 lbs). The bricks are then dried for transport to heroin processing laboratories.

Heroin synthesis is a two-step process which generally requires twelve to fourteen hours to complete. Heroin base is the intermediate product. Typically, morphine hydrochloride bricks are pulverized and the dried powder is then placed in an enamel or stainless steel rice cooking pot. The liquid acetic anhydride is then added. The pot lid is tied or clamped on, with a damp towel used for a gasket. The pot is carefully heated for about two hours, below boiling, at a constant temperature of 185 degrees Fahrenheit. It is never allowed to boil or to become so hot as to vent fumes. It is agitated by tilting and swirling until all of the morphine has dissolved. Acetic anhydride reacts with the morphine to form diacetylmorphine (heroin). This acetylation process will work either with morphine hydrochloride or *p'i-tzu* (crude morphine base). When cooking is completed, the pot is cooled and opened. The morphine and the acetic anhydride have now become chemically bonded, creating an impure form of diacetylmorphine (heroin). Water is added at three times the volume of acetic anhydride and the mixture is stirred. Activated charcoal is added and mixed by stirring and the mixture is then filtered to remove colored impurities. Solids remaining on the filter are discarded. Sodium carbonate, used at 2.5 pounds per pound of morphine, is dissolved in hot water and added slowly to the liquid until effervescence stops. This precipitates the heroin base which is then filtered and dried by heating in a steam bath for an hour. For each pound of morphine, about 11 ounces of crude heroin base is formed. The heroin base may be dried, packed and transported to a heroin refining laboratory or it may be purified further and/ or converted to heroin hydrochloride, a water-soluble salt form of heroin, at the same site.

Physical Properties

Melting point	254°C
Water of crystallization $\alpha\frac{23}{D}$	–131°C
Density	1.31
Solubility (water)	32 mg/ml
(ethanol)	64 mg/ml

Chemical Properties

Oxalic acid, sulphuric acid, and phosphoric acid, alkaloids and concentrated solutions of zinc chloride may act on morphine on two ways. Sometimes condensation takes place with the formation of compounds such morphine and tetra morphine and sometimes water is estimated according to the equation.

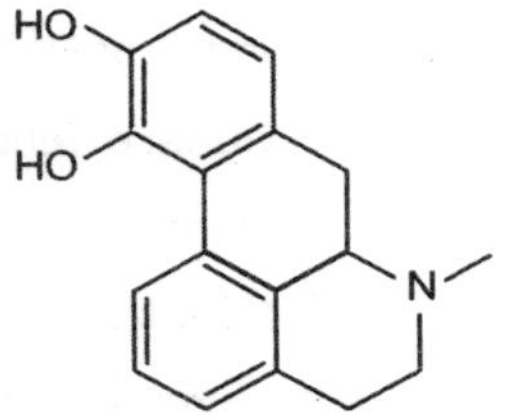

$$C_{17}H_{19}NO_3 \rightarrow H_2O + C_{17}H_{17}NO_2$$

HO

HO

N

The compound ***apomorphine*** is an amorphous, sparingly soluble base, which readily undergoes oxidation. In physiological action it differs entirely form morphine, it has no narcotic properties but is a powerful emetic.[1]

[1]Vomiting, nauseate.

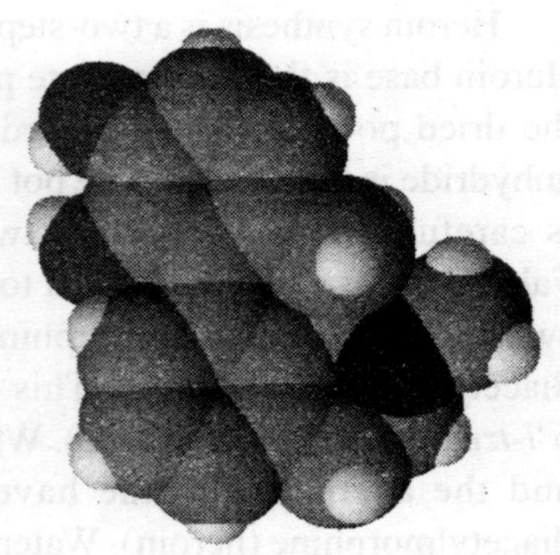

Apomorphine is a type of dopaminergic agonist, a morphine derivative (but does not actually contain morphine, or bind to opioid receptors). Apomorphine is a relatively non-selective dopamine receptor agonist, having possible slightly higher affinity for D_2-like dopamine receptors.

Historically, apomorphine has been tried for a variety of uses including psychiatric treatment of homosexuality in the early 20th century. Currently, apomorphine is used in the treatment of Parkinson's disease and (under the name **Uprima**) erectile dysfunction. It was also successfully used in the treatment of heroin addiction, a purpose for which it was championed by the author William S. Burroughs. It is a potent emetic, meaning that it should not be administered without an antiemetic such as domperidone. The emetic properties of apomorphine are exploited in veterinary medicine to induce therapeutic emesis in canines that have recently ingested toxic or foreign substances.For treatment of erectile dysfunction, it is believed that dopamine receptors in the hypothalamic region of the brain are the main target, as although dopamine receptors in the penis do facilitate erection, they do so far more weakly than those in the brain. Apomorphine is clear as a liquid but stains green. Therefore care must be taken to avoid splashes. Apormophine does not remain stable for more than 24 hours in a plastic container, so syringes are discarded if not used within 24 hours.

Physical Properties of Apomorphine

Molecular Formula	= $C_{17}H_{17}NO_2$
Formula Weight	= 267.32238
Composition	= C(76.38%) H(6.41%) N(5.24%) O(11.97%)
Molar Refractivity	= 77.86 cm^3
Molar Volume	= 205.6 cm^3
Parachor	= 568.1 cm^3
Index of Refraction	= 1.681
Surface Tension	= 58.2 dyne/cm
Density	= 1.299 g/cm^3
Polarizability	= 30.86 $10^{-24}cm^3$
Monoisotopic Mass	= 267.125929 Da
Nominal Mass	= 267 Da
Average Mass	= 267.3224 Da

Function of three Oxygen Atoms in Morphine, Relationship of Morphine to Codeine

The three oxygen atom of morphine possess different functions. On of them is resent in a phenolic hydroxyl group, which endows the alkaloids with certain acetic properties. The hydrogen of this group is replaced by metals and by alkyl radicals in codeine it is replaced by methyl group.

> Codeine is therefore regarded as the methyl ester of morphine.

This relationship between the two alkaloids was recognized by Matthiessen and Wright 1869 and represented by them as follows:

$$\underset{\text{Morphine}}{C_{17}H_{17}NO(OH)_2} \quad + \quad \underset{\text{Codeine}}{C_{17}H_{17}NO(OH)(OCH_3)}$$

By the action of concentrated hydrochloric acid on codeine at 100°C they obtained an amorphous chlorinated derivative, chlorocodite

$$C_{18}H_{21}NO_3 \quad + \quad HCl \quad \rightarrow \quad C_{18}H_{20}ClNO_2 \quad + \quad H_2O$$

When this was heated with water at 130°C codeine was regenerated with hydrochloric acid at 150, on the other hand, it was decomposed into apomorphine and methyl chloride.

$$\underset{\text{Chlorocodide}}{C_{18}H_{20}ClNO_2} \quad \rightarrow \quad \underset{\text{Apomorphine}}{C_{17}H_{17}NO_2} \quad + \quad CH_2Cl$$

On combining these two equations, it is seen that hydrochloric acid decomposes codeine at 150°C with elimination of a methyl group and a molecule of water. The solid reaction product is identical with that which results from morphine by the action of dehydrating agents. Hence codeine must be derived from morphine by the replacement of a hydroxyl group by a methyl group.

The conversion of morphine into codeine, which was effected in 1881 by Grimaux, confirmed the conclusions of Matthiessen and Wright, and definitely proved that codeine was monomethyl of morphine. Grimaux prepared codeine from morphine by treating it with methyl iodide in the presence of an alkali

$$\underset{\text{Morphine}}{C_{17}H_{17}NO(OH)_2} + CH_3I + KOH \rightarrow KI + H_2O + \underset{\text{Codeine}}{C_{17}H_{17}NO(OH)COCH_3}$$

The constitution of these two alkaloids may thus be discussed together. The second oxygen atom in morphine has been shown present in an alcoholic grouping >CH.OH since codeine on oxidation yields a ketone *codenone.*

The third oxygen atom sis very non-reactive. According to Vongerichten it is united to two carbon atoms as in ether. Roser and Howard by the use of Zeisel's method proved that the thebaine contains two methoxyl groups. Thus the relationship between the three alkaloids may be represented as follows:

				Non-reactive oxygen	Phenolic 14 2 43	alcoholic
Morphine	C17	H16	N	O	OH	HOH
Codeine	C17	H16	N	O	OCH_3	HOH
Thebaine	C17	H14	N	O	OCH_3	$HOCH_3$

Function of Nitrogen Atom in Morphine

The behaviour of morphine on *exhaustive methylation* shows that the nitrogen atom is contained in a ring and that it is attached to the three carbon atoms and therefore tertiary morphine unites directly with one molecule of methyl iodide to give a methiodide.

When methyl morphine methiodide (or codeine meth iodide) is heated with sodium hydroxide converted into a tertiary base, **methyl morphimethine.**

codeine methiodide + NaOH ⟶ methyl morphomethine + NaI

This reaction resembles the transformation of dimethyl piperidinium hydroxide into pentenyl-dimethylamine a change which necessarily involves the disruption of the piperidine ring. Hence, it may be concluded that the formation of methyl-morphimethine is due to the rupture of the nitrogen ring of the morphine.

Methyl-morphimethine exists in different isomeric forms as is a tertiary base reacting only with one molecule of methyl iodide, to give its methiodide. It constitution will be discussed more fully later.

Arrangement of the Carbon Atoms in the Morphine

Of the 17 carbon atoms in the morphine molecule 14 must belong to a phenanthrene nucleus, since the non nitrogenous decomposition products of the alkaloid have always proved to be phenanthrene derivative.

Phenanthrene was isolated by Schrötter and Vongerichten by the distillation of the alkaloid with zinc dust. Morr obtained it in the same way from methyl morphimethine

Decomposition of Morphine Codeine Thebaine

The decomposition of morphine and its derivative into nitrogenous compounds of low carbon content and nitrogen free compounds, which are enriched in carbon, may accomplished in various ways e.g.

(*i*) By the action of hydrochloric acid or acetic anhydride on metho hydroxides of morphine and codeine or methyl-morphimethine.

(*ii*) By decompositions of the ammonium bases of the morphine group under the influence of heat and alkalies.

Decomposition Products of Morphine which do not contain Nitrogen

The degradation products obtained are derivatives of the compounds morphol $C_{14}H_8(OH)_2$ described before.

The constitution of these two compounds has been established chiefly through the analytical researches of Vongerichten and the synthesis of Pschorr.

The morphol has been identified as 4:3:4-dihydroxy phenanthrene and this constitution has been confirmed by the synthesis carried out by Bayer.

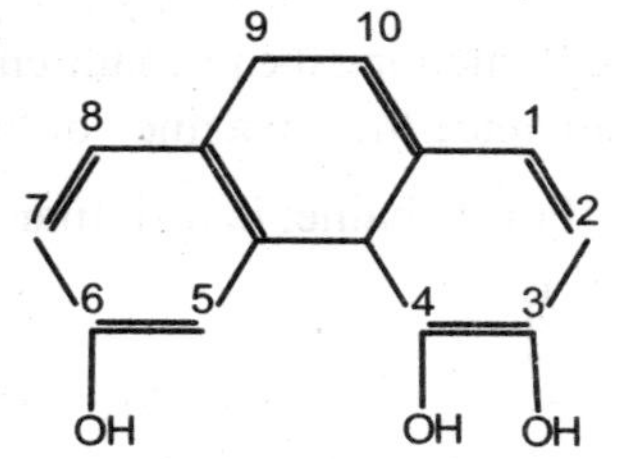

Pschorr and Vogtherr prepared the acetyl derivative of 3-methoxy 4 hydroxy-phenanthraquinone and found it to identical with the acetyl derivative of methyl morphol quinone which Vongerichten has obtained by heating methyl-morphimethine with acetic anhydride.

9 10 8 1 7 2 6 5 4 3
OH OH OH

Thus the methoxy group in codeine and the phenolic hydroxyl group in morphine each occupies position 3 in the phenanthrene nucleus. The further question as to whether the hydroxyl group in position 4 in the decomposition products corresponds to the indifferent oxygen atom or to the alcoholic hydroxyl of morphine was decided by Knorr in favour of the former assumption. He was able to show that one of the decomposition products of codenone a ketone obtained by the oxidation of the alcoholic hydroxyl groups attached to position 6. Hence the alkaloids morphine and codeine are derivative 3:4:6 trihydrooxyphenathrene.

From the investigations of thebaol and codeinone to be described later it also follows that the thebaine is derived from the same compound. The functions of the three oxygen atoms in morphine may this be indicated.

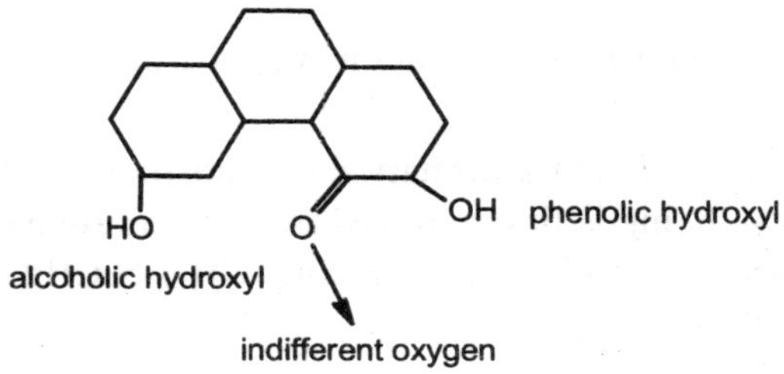

HO OH phenolic hydroxyl
alcoholic hydroxyl O
indifferent oxygen

Our knowledge of the nitrogenous decomposition of products of morphine is chiefly due to the investigation of Knorr.

The decomposition of methyl-morphimethine has given results of such importance that this compound may be considered the key to the constitution of morphine.

By the decomposition of methyl-morphimethine metho-hydroxide under the influence of heat, the volatile basic decomposition product was found to be trimethylamine, when treated with acetic anhydride the compound gave dimethylamine. Hence, of the three carbon atom which lie outside the phenanthrene nucleus in morphine, one must be united to nitrogen in the form of a methyl group.

Methyl-morphimethine was found to be decomposed by acetic anhydride to give the basic products dimethylamine and the acetyl derivative hydroxyethyl diethylamine.

$$HO{-}CH_2{-}CH_2{-}N(CH_3)_2$$

The isolation of hydroxyethyl diethylamine gave the erroneous conclusion that the methyl-morphimethine the phenathrene component and hydroxyethyldiethylamine are linked together and Knorr to advance the "oxazine" or "morpholine formula" for morphine.

For thebaine, in a similar manner, Freund obtained acetoxyethyl methyl amine

$H_3C{-}C({=}O){-}O{-}CH_2{-}CH_2{-}NH{-}CH_3$ and acetyl thebaol, this was identified by Pschorr as 3:6 dimethoxy-4-acetoxy phenathrene, thus proving that the two methoxy groups in thebaol and therefore also in thebaine occupy positions 3 and 6.

In an attempt to isolate any intermediate compound which might be formed during this decomposition Pschorr and Hass investigated the action of benzoyl chloride on the basine at 0°C. The degradation of the alkaloid was found to take place very smoothly under such conditions with the formation of the benzoyl derivatives of thebaol and hydroxyethylmethylamine.

The above oxazine or morpholine formula has to be abandoned after the discovery that the complex C-C-N could be detached from the morphine molecule *ethyl dimethyl aminoethyl ether.*

$$(H_3C)_2N{-}CH_2{-}CH_2{-}O{-}CH_2{-}CH_3$$

By heating methyl-morphimethine with sodium methoxide or thebaine and codeine methiodides with alcohols.

The above ether base however is not a primary decomposition product Knorr suggests that the three-membered chain of the side ring is first removed from the morphine alkaloids in the form of an unsaturated compounds, probably a vinyl-dimethyl amine $(CH_3)_2N.CH{:}Ch_2$, which at once combines with alcohol to give the ether base.

$$(H_3C)_2N{-}CH{=}CH_2 + H_3C{-}CH_2{-}OH \longrightarrow (H_3C)_2N{-}CH_2{-}CH_2{-}O{-}CH_2{-}CH_3$$

N,N-dimethylethenamine + ethanol → 2-ethoxy-*N,N*-dimethylethanamine

Should this prove correct, then the acetyl derivatives of the alcohol bases, obtained by the action of acetic anhydride on methyl-morphimethine, thebaine and codeine, must be regarded as secondary addition products of acetic acid with a compound which does not contain oxygen. In this case the formation of the hydramines cannot be due, as was suggested to a hydrolytic decomposition in which the "in different" oxygen of the original alkaloid is into the hydrogen group of the alcohol base.

These view have received support from the observation of Knorr and Pschorr, who found that meta-thebainone is decomposed by acetic anhydride with the formation of hydroxyethyl-dimethylamine, thus behaving similarly to methyl-morphimethine, even through it longer contains an indifferent oxygen atom.

For these reasons the assumption of an oxazine ring in morphine and thebaine has proved untenable and there is no doubt that the indifferent oxygen atom in the morphine alkaloids is present in thebaine ring, forming a bridge between position and 4 and 5 of the phenathrene nucleus. A similar structure found in methyl morphenol, which is a decomposition product of methyl-morphimethine. The complex >($C_2H_4NCH_3$) is therefore attached to the phenanthrene nucleus in methyl-morphimethine by means of carbon linking.

The same conclusion was derived by Freund from a study of the "action *of organo-magnesium halides on thebaine*"

From the cause of these reactions, Freund concluded that of three oxygen atoms in thebaine, of which two are present in methoxy groups, the third belongs to a ring similar to that occupying in diphenylene oxide. Hence the complex attached to a different position.

Knorr and Pschorr summerised their views on the constitution of the morphine alkaloids in the following statements:

(*i*) The three morphine alkaloids are derivatives of 3:6 dihydroxy – phenathrylene

(*ii*) Attached to the phenathrene nucleus as aside ring id the divalent complex –C_2H_4–N–CH_3.

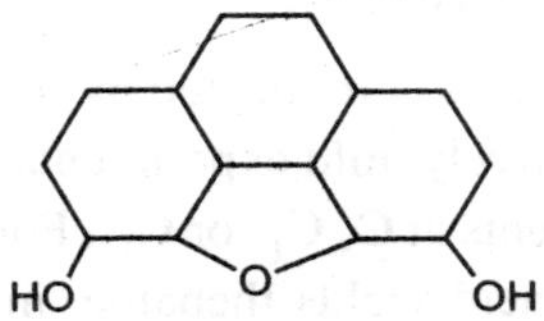

HO O OH

(*iii*) The nucleus present in thebaine is tetrahydro-phenathrene, whilst that in morphine and codeine is hexahydro-phenathrene. The six additional hydrogen atoms in morphine are distributed between rings II and III, where as ring I to which the phenolic hydroxyl group is attached, possesses true aromatic properties. From the course of the degradation it appears that the complex – C_2H_4 – N – CH_3 belongs to the reduced part of the phenathrene nucleus.

A concentration all the available facts led Knorr to put forward a formula fro morphine in arrangement with that advanced by Fallis, a year earlier and which differed from formula I in having the double bond at 8:14 instead of 7:8 and the ethanamine chain attached to C_5 in place of C_{13}.

Although these formulæ satisfy most of the experimental facts, there are a number of points which they fail to explain.

The points still at issue are the position of the double bond and the mode of attachment of the carbon end of the C – C – N chain to the phenathrene nucleus.

In 1925 it was suggested independently by Gulland and Robinson, and by Weiland and Kotake that double bond should be allocated to 7:8 instead of 8: 14 as in Knorr's formula.

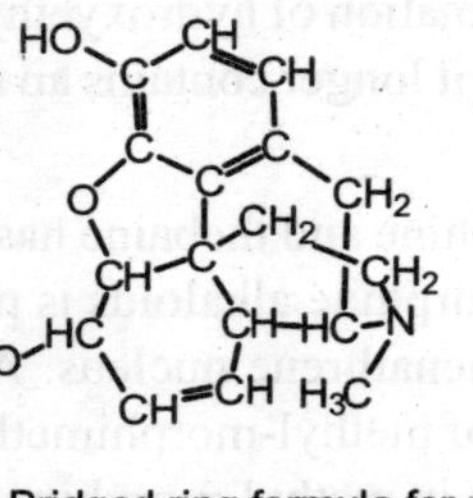

Bridged ring formula for morphine

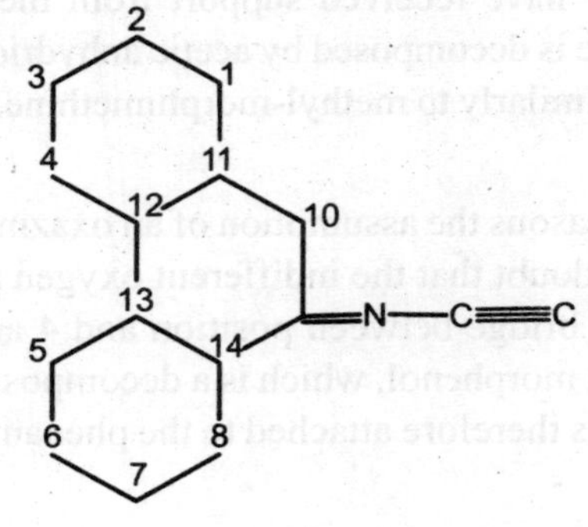

Further confirmation of this new arrangement was given by the investigation carried out by van Duin, Robinson and Smith on *neopine*, a comparatively rare alkaloid found in opium which has been shown to be an isomer of codeine having the double bond as in the 8:14 position. In 1923 Gulland and Robinson advanced argument in fabour of a C_{13} or C_{14} linkage for the ethanamine side chain, a modification which is supported indirectly by Schopf's proof that the union is not in the C_5 position as also by the direct experimental work of Wieland and Small, Gulland and Robinson's formula for morphine is now generally accepted.

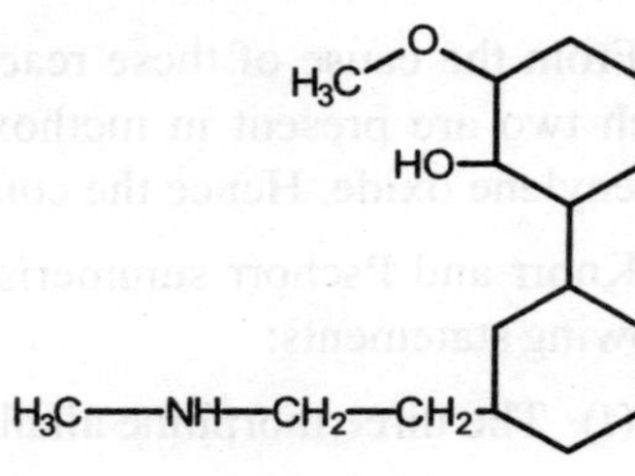

The final choice between C_{13} and C_{14} is complicated by the surprising migration which may occur in the alkaloids of this group under relatively mild experimental conditions, leading to the formation of derivation products containing substituents at C_5 C_{13} or C_{14}. For example, thebaine on being heated for a short time with dilute hydrochloric acid yields thebaine, in which the C – C – N chain is found attached to the 5th position. On the whole, however, the evidence indicates C_{13} as the most portable point of attachment, as in formula II.

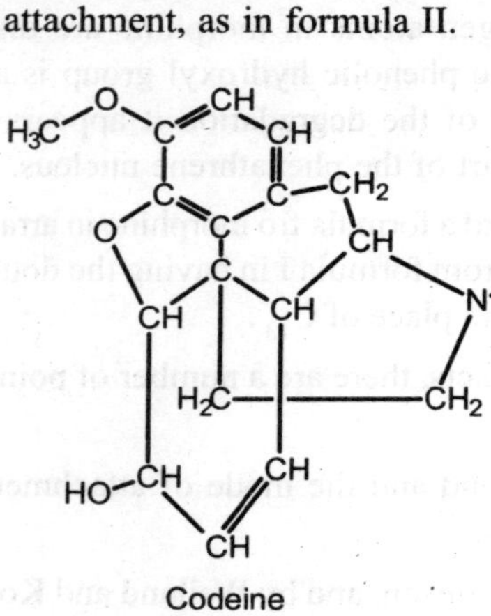

Codeine

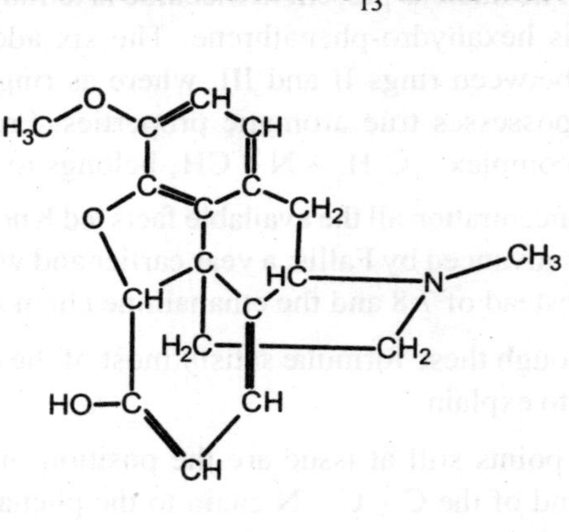

Thebaine

Thebaine is a less hydrogenated alkaloid represented by Gulland and Robinson as having formula III identical with that of Knorr except for the different point of union of the C – C – N chain. The formula supported by Schöpfs reduction of thebaine to tetrahydrothebaine.

Conversion of Thebaine to Codaine

As it is already been stated thebaine resembles morphine and codeine in constitution, but represents a different degree of hydrogenation of the phenanthrene nucleus. Attempts were therefore made to establish as experimental connection between these two structures.

Morphine

CH_3I +KOH

Codeine

oxidation

reduction

Codenone

debromination(CH_3Br)
eliminated

hydrolysis
or by bromine

Thebaine

This was carefully accomplished by Knorr through the compound codeinone, obtained by oxidation codeine with chromic acid or potassium permanganate. The changes involved are summerised in the forgoing scheme, in which the heavy type denotes groups taking parts in the reactions.

It is apparent from the above example and formulæ, codeinone is the ketone corresponding to the alcohol codeine. It is also closely related to thebaine of codeinone. In this respect thebaine is related to codeinone in the same way as codeine is to morphine. It was therefore of interest to discover a means of converting thebaine into codeinone and codeinone into thebaine.

The first part of the problem has already been solved in two ways as indicated above. On the otherhand Knorr succeded in converting thebaine into codeinone by simple hydrolysis with hot or cold acids and on the other hand, Freund observed the formation of codeinone from the bromo-derivative of thebaine.

Alakloids of Medow Saffron

Colchicine is a highly poisonous natural product and secondary metabolite, originally extracted from plants of the genus *Colchicum* (Autumn crocus, also known as the "Meadow saffron"). Originally used to treat rheumatic complaints and especially gout, it was also prescribed for its cathartic and emetic effects. Its present medicinal use is mainly in the treatment of gout; as well, it is being investigated for its potential use as an anti-cancer drug. It can also be used as initial treatment for pericarditis and preventing recurrences of the condition. In neurons, axoplasmic transport is disrupted by colchicine.

CHAPTER

37

The Organometallic Compounds

Introduction

The organo-metallic compounds are usually prepared by the action of metals such as zinc, magnesium or mercury on alkyl iodides, owing to their reactivity are frequently employed in synthetic reactions. The zinc and magnesium compounds are most important of this class have been carefully studied.

Organo-Lithium Compounds

Lithium

Of all the organo-alkali compounds those of lithium, RLi, are the most useful in synthetic organic chemistry. This is true not only because these compounds are safe and easy to manipulate in the laboratory but because they are more reactive than the Grignard reagent, and accordingly will undergo certain reactions that the Grignard's reagent will not. Despite a rapid decline in recent years, however, the price of lithium metal is still too high to permit a wide use of these organo metallic compounds on an industrial scale.

Physical Properties

Organolithium compounds are either salt like solids or liquids. In the straight-chain aliphatic series, methyl- and ethyl-lithium are colourless solids whereas the higher members are liquids. In the aromatic series, the compounds are usually colorless solids; if the metal, however, is bonded to a carbon atom which in turn is joined to one or more aromatic nuclei, the compounds are generally highly coloured. Thus, benzyllithium is yellow, and triphenylmethyllithium is red. The carbon-metal bond in such compounds is usually considered ionic, and the color of these compounds is undoubtedly associated with resonance possibilities in the anion.

Reactions

Like the Grignard reagent, organolithium compounds react with active hydrogen, carbonyl groups, nitriles, carbon dioxide, and oxygen:

$$RLi + R'OH \rightarrow RH + R'OLi$$

$$RLi + R'CHO \rightarrow RCH(OH)R'$$

$$RLi + R_2'CO \rightarrow RR_2'COH$$

$$RLi + R'COOR' \rightarrow RR_2'COH$$

$$RLi + R'CN \rightarrow RR'CO$$

$$RLi + CO_2 \rightarrow RCOOH$$

$$RLi + O_2 \rightarrow ROH$$

In all of these reactions the lithium reagent may be either aliphatic or aromatic. In addition to the general reactions above, organolithium reagents are able to undergo certain special reactions which the Grignard reagent either does not undergo or does so at such a slow rate as to be relatively useless. These include metalations (metal-hydrogen interchange), halogen-metal interconversion, anil additions, and additions to olefinic linkages. The following equations give two specific examples of each of these reaction types:

$$\text{N–CH}_3 + C_6H_5Li \longrightarrow \text{N–}CH_2^- Li^+ + C_6H_6$$

$$\text{S} + n\text{-}C_4H_9Li \longrightarrow \text{S–Li} + n\text{-}C_4H_{10}\uparrow$$

$$\text{Br–C}_6\text{H}_4\text{–Cl} + n\text{-}C_4H_9Li \longrightarrow \text{Li–C}_6\text{H}_4\text{–Cl} + n\text{-}C_4H_9Br$$

$$\text{Br} + n\text{-}C_3H_7Li \longrightarrow \text{Li} + n\text{-}C_3H_7Br$$

$$\text{N} + C_6H_5Li \longrightarrow \text{N–}C_6H_5$$

Cl, Cl, N + Li–(C6H4)–OCH3 → Cl, Cl, N, OCH3

+ C_6H_5Li → Li; H_3C—C—CH_3; C_6H_5

$$(C_6H_5)_2C{=}CH_2 + n\text{-}C_4H_9Li \xrightarrow{CO_2} (C_6H_5)_2C(COOH)CH_2(n\text{-}C_4H_9)$$

Preparation

In general, organolithium compounds are best prepared by the action of alkyl or aryl halides upon lithium metal in a relatively inert solvent, like ethyl ether. All precautions observed in the preparation of the Grignard reagent must be observed, such as rigorous exclusion of moisture and oxygen. Usually this is accom-plished by the use of an inert atmosphere like nitrogen, butane, or helium.

Since most aliphatic organolithium compounds (methyilithium excepted) will gradually cleave ethyl ether, petroleum ether fractions are sometimes used as solvents. However, this cleavage can often be minimized by short reaction times and low temperatures. The following directions for the preparation of phenyllithium and n-butyllithium are given as typical of the aromatic and aliphatic types.

Phenyllithium, C_6H_5Li

In a dry 500-ml. three-necked flask fitted with a dropping funnel, mechanical stirrer, and reflux condenser protected from moisture, the whole being swept with dry nitrogen, are placed 3.5 gram (0.5 gram-atom) of lithium cut into small pieces and 100 ml. of dry ether. The stirrer is started, and about 10 ml. of a solution of bromobenzene (0.25 mole) in 50 ml of dry ether is added from the dropping funnel; a vigorous reaction usually takes place. The remainder of the solution is added gradually over a half-hour period, when the metal should have largely disappeared. The yield is 90% or better.

n-Butyllithium, C_4HgLi

Into a 500-ml. three-necked flask equipped with a stirrer, low-temperature thermometer, and a dropping funnel are placed 200 ml. of anhydrous ether (dried over sodium). After sweeping the apparatus with dry nitrogen, 8.4 grams (1.2 gram-atoms) of lithium wire is wound in a loose coil and rinsed with ether, and the flattened coil cut into pieces about 7 mm. in length which are allowed to fall directly into the reaction flask in a stream of nitrogen. With the stirrer started, about 30 drops of a solution of 68.5 grams (0.5 mole) of n-butyl bromide in 100 ml. of ether are added from a dropping funnel, and the reaction mixture is then cooled to –10°C. With a dry ice-acetone bath kept at approximately -30 to -

40°C. The remainder of the *n*-butyl bromide is added at an even rate over a half-hour period while maintaining the internal temperature at –10°C. After addition is complete, the reaction mixture is allowed to warm up to 0-10°C. While being stirred for one to two hours. The reaction mixture is then filtered by decantation through a narrow tube plugged with glass wool into a graduate dropping funnel previously flushed with nitrogen. The yield after filtration is about 80-85%.

Organo Sodium Compounds

For example sodium methyl, $NaCH_3$ are obtained by the action of sodium on the corresponding mercury alkyls. In the pure state they form colourless amorphous which are completely insoluble in different solvents, and when heated decomposes without melting.

These are very sensitive towards oxygen, moisture and carbon dioxide, are inflammable in air and very reactive.

Preparation

A general method for the preparation of both aromatic and ali-phatic organo sodium compounds involves the reaction of sodium metal on the corre-sponding zinc or mercury compound:

$$R_2Hg + 3\,Na \rightarrow NaHg + 2\,RNa$$
$$R_2Zn + 3\,Na \rightarrow NaZn + 2\,RNa$$

Although these reactions usually result in good yields, they have a disadvantage in that organo zinc compounds are often spontaneously flammable, and organo-mercury compounds (especially the lower alkyl series) are highly poisonous. However, organo-sodium compounds can sometimes be prepared directly from sodium and a variety of halides. Thus, with sodium amalgam, chlorobenzene in benzene gives phenyl sodium in 89% yield, n-amyl chloride yields *n*-amyl sodium, and triphenylmethyl chloride in ether forms triphenylmethylsodium. Phenyl sodium has also been prepared by dropping chlorobenzene into a stirred suspension of finely dispersed sodium in benzene. Occasionally a metal-hydrogen interchange has been used to prepare an aromatic sodium compound. Thus toluene reacts with ethylsodium to form benzylsodium and ethane.

Disodium compounds can sometimes be prepared by adding sodium directly to an olefinic linkage. Thus tetraphenylethylene adds sodium easily to form tetraphenyl-ethylenedisodium:

$$(C_6H_5)_2C = C(C_6H_5)_2 + 2\,Na \rightarrow (C_6H_5)_2C — (C_6H_5)_2C\ Na_2^{2+}$$

All of the above preparations must be carried out under rigorously anhydrous conditions and in an inert atmosphere. The following directions for the preparation of phenyl- and amyl-sodium are given as typical of aromatic and aliphatic types.

Phenylsodium, C_6H_5Na

To 27 grams (1.17 gram-atoms) of sodium, freshly dispersed in 150 ml of toluene, is added 10-15 ml of a solution of 112 grams (1 mole) of chlorobenzene in 112 grams of toluene. If the temperature does not rise rapidly during the first ten minutes of mechanical agitation, 2 ml. of amyl alcohol is added to initiate the reaction. Once the reaction has started, the remainder of the chlorobenzene solution is added at a rapid dropwise rate over a twenty-minute period. A cooling bath is required to maintain the internal temperature below 30°C. during this addition. Carbonation of the reaction mixture gives a 99% yield of benzoic acid. If the reaction mixture is refluxed for two hours before carbonation on dry ice, a 97% yield of phenylacetic acid is obtained.

n-Amylsodium, $C_5H_{11}Na$

A mixture of 11.5 grams (0.5 gram-atom) of sodium and 340 ml of *n*-octane are placed in a Morton flask and then heated to 105°C. The mixture is stirred at this temperature with a stirrer operating at 10,000 r.p.m. until the metal is finely dispersed. The stirrer is stopped, the flask is cooled, and 30.2 ml (0.25 mole) of n-amyl chloride diluted with 30 ml. of *n*-octane is added over a period of one hour, during which time a bath of kerosene and solid carbon dioxide keeps the temperature at 0°C. After one additional hour of stirring (all at 10,000 r.p.m.) at this temperature, the yield is shown by carbonation to be 72%.

Potassium

Organopotassium compounds, RK, being more reactive than the corresponding sodium compounds, are rather difficult and dangerous to handle. Because of their great basic strength they are valuable as metalating agents in organic syntheses.

Physical Properties

These compounds are electrovalent, salt like, and generally insoluble in most solvents with which they do not react.

Reactions

Organo-potassium compounds undergo all the reactions exhibited by the sodium compoynds. At the moment organo-potassium compounds are used principally in small-scale reactions involving metal-hydrogen interchange. Thus, cumene reacts with ethyl potassium to form phenylisopropylpotassium:

$$C_6H_5CH(CH3)_2 + C_2H_5K^+ \rightarrow C_6H_5C(CH_3)_2\,K^+ + C_2H_6$$

Phenylisopropylpotassium, in turn, can be used to metalate diphenyl- and triphenyl-methane on the central carbon atom. A host of other reactions of this general acid base type are possible with this and other potassium compounds.

Preparation

Aliphatic potassium compounds are usually prepared by the reaction of the metal with the corresponding mercury or zinc compounds. The aromatic types are generally made by metal-hydrogen interchanges. A very useful method for preparing phenylisopropylpotassium was developed by Ziegler and may be used for related types. This involves the cleavage of phenylisopropyl methyl ether with sodium-potassium alloy:

$$C_6H_5C(CH_3)_2OCH_3 + NaK \rightarrow C_6H_5C(CH_3)_2\,K^+ + CH_3O^-K^+$$

Phenylisopropylpotassium

Phenylisopropylpotassium is perhaps one of the most useful of the organo potassium compounds, since it is not as difficult to handle as the simple alkyl types and is a very effective metalating agent.

Potassium does not add readily to olefinic linkages as sodium does, probably because of steric hindrance. Sodium compounds can often be converted to the corresponding potassium compounds by treatment with potassium metal. This is in keeping with the general rule that the more reactive organometallic compound can be prepared from one of lesser reactivity.

Because of its importance in this series, directions for the preparation of phenyl-isopropylpotassium are listed below:

Phenylisopropylpotassium, $C_6H_5C(CH_3)_2$ K

Fifteen grams (0.10 mole) of phenylisopropyl methyl ether is dissolved in 700 ml of anhydrous ether, and 12 ml of sodium-potassium alloy (1: 5 by weight) is added. The mixture is stirred for 10 hours under nitrogen, during which time the solution turns dark red. It is filtered under nitrogen into a graduated dropping funnel, where titration with standard acid shows the yield to be about 80 per cent.

Zinc Alkyls

These compounds are discovered in 1849 by Frankland. They are obtained by the action of excess of zinc in the form of the zinc-copper couple and by addition of ethyl acetate. Zinc alkyl iodides are first formed which decomposed into zinc alkyls and zinc iodide on distillation.

$$C_2H_5I \quad + \quad Zn \quad \rightarrow \quad C_2H_5ZnI$$

$$C_2H_5ZnI \rightarrow Zn(C_2H_5)_2 \; + \; ZnCl_2$$

The zinc alkyls are colourless, unpleasent smelling liquids, which boil without decomposition at relatively low temperature in an atmosphere of carbon dioxide. They are spontaneously inflammable in air, and produce painful burns in contact with the skin. Consequently they must be handled with caution.

Neverthless they were examined in detail by Frankland, who showed that they could be used for the synthesis of a variety of compounds, including alcohols and ketones.

$Zn(CH_3)_2$ — zinc methyl $\qquad$ $Zn(CH_2-CH_3)_2$ — zinc ethyl

Paraffins are also produced on heating zinc alkyls to a high temperature with alkyl iodides.

$$Zn(CH_3)_2 + 2CH_3I \rightarrow ZnI_2 + 2CH_3{-}CH_3$$

Frankland's work on the metallic was extended to derivatives of other metals e.g. trimethyl; arsine and $As(CH_3)_3$ trimethyl stebine $Sb(CH_3)_3$, which first led to brief that each element has a definite combining power and so laid to the foundation of the modern theory of valency.

Synthesis of Cyclopropanes using RZnX (The Simmons-Smith reaction).

$$I\text{-}CH_2\text{-}I \quad Zn \xrightarrow[Et_2O]{Cu} I\text{-}CH_2\text{-}ZnI$$

$$I\text{-}CH_2\text{-}ZnI \quad CH_2{=}CH\text{-}R \xrightarrow{Et_2O} \text{cyclopropane (}CH_2\text{, R)} \quad Zn\,I_2$$

Reaction type: 1. Oxidation-Reduction, 2. Addition

Summary

- This is the most important reaction involving an organozinc reagent.
- Also known as the Simmons-Smith reaction
- The iodomethyl zinc iodide is usually prepared using Zn activated with Cu.
- The iodomethyl zinc iodide reacts with an alkene to give a cyclopropane.
- The reaction is stereospecific with respect to to the alkene (mechanism is concerted).

Substituents that are *trans* in the alkene are *trans* in the cyclopropane *etc.*

$$I\text{-}CH_2\text{-}ZnI \quad \text{trans-}CH_3CH{=}CHCH_3 \xrightarrow{Et_2O} \text{trans-1,2-dimethylcyclopropane} \quad Zn\,I_2$$

$$I\text{-}CH_2\text{-}ZnI \quad \text{cis-}CH_3CH{=}CHCH_3 \xrightarrow{Et_2O} \text{cis-1,2-dimethylcyclopropane} \quad Zn\,I_2$$

$$\text{alkene} + I\text{-}CH_2\text{-}ZnI \longrightarrow \text{cyclopropane (}CH_2\text{)} \quad ZnI_2$$

Mechanism of the Simmons-Smith Reaction

A concerted reaction : both new **C-C** are formed simultaneously. Best viewed as the nucleophilic **C=C** causing loss of the iodide leaving group and the electrons from the nucleophilic **C-Zn** bond being used to form the other **C-C** bond.

Zinc

Although diethylzinc was the first organometallic compound prepared (in 1849) and organozinc compounds were widely used as intermediates in organic syntheses for half a century, the popularity of the zinc compounds declined immediately after Grignard's announcement of his new magnesium reagent in 1900. Unlike the magnesium compounds, the R_2Zn compounds are better known than the RZnX.

	B.p. °C	Density	Refrative index
$(CH_3)_2Zn$	46_{760}	$1.386^{10.5}_{4}$	—
$(C_2H_5)_2Zn$	118_{760}; 30_{27}	1.2065^{20}_{4}; 1.245^{8}_{4}	—
$(CH_3CH_2CH_2)_2Zn$	160_{760}; 48_{10}	$1.1034^{20.2}_{4}$	$1.4845^{18.6}_{D}$
$(CH_3)_2CHCH_2)_2Zn$	170_{760}; 55_{10}	$1.0080^{16.5}_{4}$	1.4603^{16}_{D}

The diarylzinc compounds are white crystalline solids. Thus, diphenylzinc forms fine white needles, soluble in ether and benzene and melting at 107°C. *Bis*(p-chlorophenyl)zinc crystallizes in small needles which melt at 212-214°C. *Bis*(p-dimethyl aminophenyl)zinc melts at 135-137°C. and is insoluble in petroleum ether.

Alkylzinc iodides, when pure, are white crystalline materials which usually decompose upon heating to the symmetrical compound:

$$2\ RZnI \rightarrow R_2Zn + ZnI_2$$

Reactions

Low-molecular-weight zinc alkyls are spontaneously flammable in air, burning vigorously with a blue flame and giving off dense white fumes of zinc oxide. Generally speaking, the higher-molecular-weight zinc alkyls and zinc aryls are not spontaneously flammable, but nevertheless do undergo oxidation quite readily.

Unlike the Grignard reagents, zinc compounds do not react with carbon dioxide, and consequently this gas, as well as nitrogen, helium, etc., can be used as an inert atmosphere during the preparation of the zinc compounds.

Like any other reactive organometallic compound, zinc compounds will react vigorously with active hydrogens:

$$R_2Zn + 2\ H_2O \rightarrow 2\ RH + Zn(OH)_2$$

$$R_2Zn + 2\ HX \rightarrow 2\ RH + ZnX_2$$

$$R_2Zn + 2\ ROH \rightarrow 2\ RH + Zn(OR)_2$$

$$R_2Zn + 2\ NH_3 \rightarrow 2\ RH + Zn(NH_2)_2$$

In addition to these rather commonplace reactions, there are three types of syntheses in which organozinc compounds are still utilized today. The first and perhaps most important is the *Reformatsky reaction* in which a halogenated ester is treated simultaneously (similarly to the Barbier technique) with zinc and either an aldehyde or a ketone to form either a secondary or a tertiary alcohol:

$$BrCH_2COOC_2H_5 + Zn + RCHO \rightarrow RCH(OH)CH_2COOC_2H_5$$

$$CH_3CHBrCOOC_2H_5 + Zn + RCOR' \rightarrow RR'C(OH)CH(CH_3)COOC_2H_5$$

The Grignard reagent or organoalkali compounds cannot be used in this type of reaction since, because of their greater reactivity, they will attack the ester linkage.

A similar situation prevails when at low temperatures organo zinc halides couple with acid chlorides to form ketones rather than tertiary carbinols:

$$RZnX + RCOCl \rightarrow RCOR + ZnXCl$$

In this case advantage is taken of the relatively slow reaction of the zinc compound with the ketone at low temperatures. However, this use for the zinc compounds has declined since Gihnan and Nelson found that organo-cadmium compounds are better for ketone syntheses. Even the Grignard reagent can be used in this type of reaction if certain precautions are observed.

Organozinc compounds are still used today as intermediates in the preparation of more reactive organometallic compounds. Thus, when they are treated with an alkali or alkaline earth metal, the zinc is usually replaced in good yields:

$$R_2Zn + Ca \rightarrow R_2Ca + Zn$$

$$R_2Zn + 2\ Na... \rightarrow 2\ RNa + Zn$$

Preparation

Primary alkyl zinc iodides are prepared by heating together the alkyl iodide with copper-zinc alloy in toluene, benzene, or xylene, together with some anhydrous ethyl acetate. The reaction is usually started by the addition of some iodine, and the yields are generally in the neighborhood of 80 per cent.

Secondary alkyl zinc iodides are prepared in a similar fashion in about 60% yields, using petroleum ether as a solvent. Tertiary alkylzinc iodides usually are not formed in good yields.

Arylzinc halides are usually prepared with the Grignard reagent of the aryl halide:

$$RMgX + ZnCl_2 \rightarrow RZnX + MgCl_2$$

Zinc dialkyls can be formed by treating alkyl iodides with zinc metal. This reaction presumably involves the intermediate formation of the alkylzinc iodide, which in turn decomposes into dialkylzinc:

$$Zn + RI \rightarrow RZnI$$

$$2\ RZnI \sim ZnI2 + R_2Zn$$

Both the dialkyl and the diaryl compounds are often best prepared by the Grignard reaction:

$$ZnCl_2 + 2RMgX \rightarrow R_2Zn + MgCl_2 + MgX_2$$

Diethylzinc, $(C_2H_5)_2Zn$

In a 500-ml round-bottomed flask provided with a reflux condenser and a heavy stirrer is placed 130 grams (2 gram-atoms) of zinc-copper couple. To this is added a mixture of 78 grams (0.5 mole) of ethyl iodide and 54.4 grams (0.5 mole) of ethyl bromide; the stirrer is started, and the mixture is heated to reflux. After one-half hour of refluxing the reaction starts, as evidenced by the greatly increased rate of refluxing, and the flame is removed. If the reaction be-comes too vigorous, the flask is cooled with ice water, but only to the point when the reaction is again under control. At the end of one-half hour from the time the flame is removed, the reaction is usually over. The flask is allowed to cool, and collected with a distilling head, condenser, and receiver, and the contents are distilled under reduced pressure directly from the reaction flask into a 200-ml round-bottomed flask immersed in an ice-salt mixture. At the end of the distillation dry carbon dioxide or purified nitrogen is admitted to the apparatus. The yield of crude material, which is sufficiently pure for most purposes, is 53-55 grams (86-89%).

Organo-Magnesium Compounds

It is only in modern times with the discovery that free magnesium alkyls could be replaced by the readily soluble compounds of the type RMgI, that the organo magnesium derivatives have been used with such great success in synthesis.

Grignard Reagent

Victor Grignard was an enthusiastic young French chemist who discovered how to make organo magnesium halides (RMgX) while working for his Ph.D. His boss, Barbier, had been trying this sort of chemistry for some time, but Victor was the genius who solved the problem. This discovery in 1901 changed the course of organic chemistry and won him the Nobel Prize in 1912. We now refer to such compounds as **Grignard Reagents** - and they are the first tool in your bag. Victor's breakthrough came with two discoveries - an *ether solvent* was vital and the whole chemistry must be carried out *bone dry*. You take an alkyl halide (preferably a bromide or iodide but a very reactive chloride such as tertiary-butyl chloride or benzyl chloride will be OK) magnesium metal and ether (dried with sodium metal) and with a little persuasion you get a vigorous reaction resulting in a Grignard reagent.

$$MeCH_2Br + Mg \xrightarrow[\text{ether}]{\text{dry}} MeCH_2 * Mg * Br \qquad \left[\begin{array}{c} OEt_2 \\ \downarrow \\ MeCH_2\text{-}Mg\text{-}Br \\ \uparrow \\ OEt_2 \end{array} \right]$$

The Grignard Reaction — How the ether is involved

The reagent is used in solution (it is soluble in ether) and is never isolated. The metal is positively polarised and the alkyl group thus is like a carbanion. It certainly behaves as a carbanion. All kinds of alkyl halides react - and amazingly, even bromobenzene and other aryl bromides and iodides react easily with magnesium. This is particularly surprising since the aromatic halogen is so unreactive.

Reactions of Grignards

They react vigorously with compounds with 'active' hydrogens – OH, NH and others. That's why the solution must be dry.

$$MeCH_2 * Mg * Br \xrightarrow{H_2O} MeCH_2\text{-}H + HO * Mg * Br$$

$$\xrightarrow{ROH} MeCH_2\text{-}H + RO\text{-}Mg * Br$$

Grignards react enthusiastically with all kinds of C=O bonds.

$$\overset{\delta^-}{R}-\overset{\delta^+}{MgBr} + \;>C=O \longrightarrow R-\underset{|}{\overset{|}{C}}-OMgBr$$

The 'carbanionic' R group attacks the electron-poor carbonyl carbon. This is a very useful reaction. For example: **reaction with carbon dioxide:**

$$R-Br \xrightarrow{Mg} \overset{\delta^-}{R}-\overset{\delta^+}{MgBr} + O=C=O \longrightarrow R-\overset{O}{\overset{\|}{C}}-OMgBr \xrightarrow{H^+} R-\overset{O}{\overset{\|}{C}}-OH$$

Making a carboxylic acid from an alkyl bromide.

This is a fantastically good way of making carboxylic acids. After the reaction of the Grignard with carbon dioxide, addition of mineral acid liberates the carboxylic acid. So we can convert an alkyl bromide to an acid - our first **C-C** bond-forming reaction.

It was also shown by Tshellinzeff that the formation of these compounds also takes place slowly in other solvents such as benzene, toluene and xylene, in the presence of a trace of ether. The amount of organo magnesium halide formed is out of all proportion to the quantity of ether employed from which it was concluded that in the Grignard Reagent part of a catalyst.

Tshellinzeff also found that other substance could act as catalysts the formation of organo-magnesium compounds takes place in solvents such as benzene, toluene xylene and petroleum ether, when a few drops of a tertiary amine (e.g. dimethyl aniline) added; In this case the magnesium compound is discarded out of the solution as a white precipitate corresponding to the formula R – Mg – X. This is sometimes of practical as well as theoretical importance, since the catalytic influence of the tertiary amine is in some cases far more energetic than that of the tertiary amine is in some cases far more energetic than that of ether and the reaction often takes place more rapidly and with as good as a yield as by the Grignard's Method. Nevertheless, Grignard's method of preparing the compounds in dry ether solution is generally more convenient. If desired, the magnesium alkyl halides may be isolated in combination with two ether molecules e.g $CH_3MgI\ 2(C_2H_5)_2O$.

According to Meisenheimer these compounds are regarded as complexes of magnesium in which the metal occurs as the central atom with a coordination number 4.

$$\begin{array}{l} H_3C-CH_2-CH_2-CH_2-O \diagdown \quad\quad \diagup CH_3 \\ \qquad\qquad\qquad\qquad\qquad\qquad Mg^{2-} \\ H_3C-CH_2-CH_2-CH_2-O \diagup \quad\quad \diagdown I \end{array}$$

By reason of their extraordinary reactivity the organo-magnesium compounds aroused great interest and the original investigations of Grignard where immediately followed by those of a number of other worker. It was soon apparent that for synthetic purposes magnesium compounds of the type R – Mg – X were more conveniently manipulated gave better yields, and were of more general utility than the zinc alkyls. Already they have attained a supreme position in synthetic chemistry which is unparallel to that of any other class of compound. A recent development is the discovery by Schlenk that the organo-magnesium halide and the compound of type MgR_2

$$2CH_3.MgI \leftrightarrows Mg(CH_3)_2 + MgI_2$$

It therefore appears probable that some of the activity of the Grignard reagent is associated with the presence of magnesium dialkyl or diaryl. With the aid of the Grignard reaction it is possible to synthesize hydrocarbon primary, secondary and tertiary alcohols, ethers, ketones aldehydes carboxy and thio-acids, phenols and thiophenols, and a variety of nitrogen compounds, as well as other alkyl metallic derivatives.

For example of the use of alkyl magnesium halides in synthetic work.

Experiment: Add 3 g of clean magnesium ribbon or fillings to 75cc of perfectly dry ether and then pour in 18 g of methyl iodide, the flask being attached with reflux condenser. The magnesium rapidly dissolves. Cool add to the solution 7 g of acetone, a white bulky precipitate of magnesium compound is precipitated. On dissolved in sulphuric acid in dilute state, the ether solution of the alcohol separates, and may be withdrawn and distilled.

ARSENO ORGANIC COMPOUNDS

Introduction

Mainly in sequence of the researches of P . Ehrlich organic compounds of arsenic have been extensively employed in medicine. Only a few of the more important of these will treated here.

Cacodyl Oxide

Cacodyl oxide is a chemical compound of the formula $[(CH_3)_2As]_2O$. This organoarsenic compound is primarily of historical significance as it is sometimes considered to be the first organometallic compound synthesized in relatively pure form. "Cadet's fuming liquid", which is composed of cacodyl and cacodyl oxide, was originally synthesized by heating sodium acetate with arsenic trioxide. It has a disagreeable odor and is toxic. It has been used as a denaturing and warning agent.

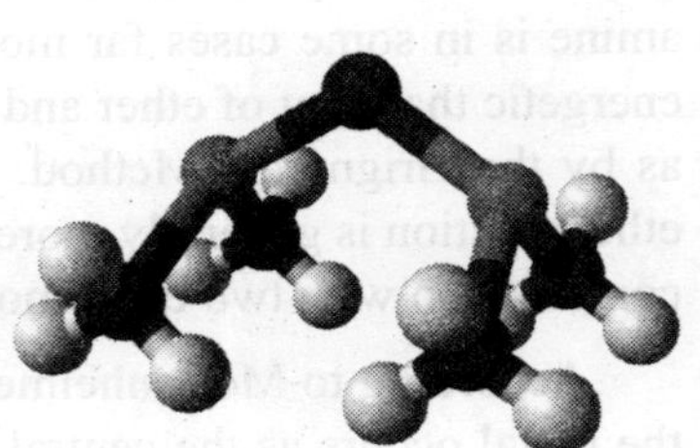

Cacodylic Acid

Significant early research into cacodyls was done by R. W. Bunsen at the University of Marburg. Bunsen said of the compounds, "the smell of this body produces instantaneous tingling of the hands and feet, and even giddiness and insensibility . . . It is remarkable that when one is exposed to the smell of these compounds the tongue becomes covered with a black coating, even when no further evil effects are noticeable". His work in this field led to an increased understanding of the methyl radical.

Cacodyloxide, $(CH_3)_2As)_2O$, is often considered the first organometallic compound to be prepared synthetically.

Cacodylic acid is the chemical compound with the formula $(CH_3)_2AsO_2H$. Derivatives of cacodylic acid, **cacodylates**, were frequently used as herbicides. For example, "Agent Blue," one of the defoliants in the Vietnam War, was a mixture of cacodylic acid and sodium cacodylate.

Synthesis and Reactions

In the 1700s it was known that combining As_2O_3 and four equivalents of potassium acetate (CH_3CO_2K) gives a product called "Cadet's fuming liquid" which contains cacodyl oxide, $((CH_3)_2As)_2O$.

Cacodylic acid can be reduced to dimethylarsine (III) derivatives, which are versatile intermediates for the synthesis of other organoarsenic compounds

$$(CH_3)_2AsO_2H + 2\ Zn + 4\ HCl \rightarrow (CH_3)_2AsH + 2\ ZnCl_2 + 2\ H_2O$$

$$(CH_3)_2AsO_2H + SO_2 + HI \rightarrow (CH_3)_2AsI + SO_3 + H_2O$$

Physcial Properties

Molecular Formula	$= C_2H_7AsO_2$
Formula Weight	= 137.99738
Composition	= C(17.41%) H(5.11%) As(54.29%) O(23.19%)
Density	=> 1.1 g/cm3
Melting point	=192 - 198 °C
Boiling point	= > 200 °C
Solubility in water	667 g/l
Monoisotopic Mass	= 137.966201 Da
Nominal Mass	= 138 Da
Average Mass	= 137.9974 Da

Health Effects

Cacodylic acid is highly toxic by ingestion, inhalation, or skin contact. Once thought to be a byproduct of inorganic arsenic detoxification, it is now believed to have serious health consequences of its own. It has been shown to be teratogenic in rodents, most often causing cleft palate but also fetal fatality at high doses. It has been shown to be genotoxic in human cells, causing apoptosis and also decreased DNA production and shorter DNA strands. While not itself a strong carcinogen, cacodylic acid does promote tumors in the presence of carcinogens in organs such as the kidneys and liver.

Sodium cacodylate is frequently used as a buffering agent in the preparation and fixation of biological samples for transmission electron microscopy.

Primary Arsonic (arsenic) Acids

It was shown by Elrlich and Berthman that the arsenic compound discovered in 1863 by Bechamps by heating aniline aresenate and employed in medicine under the name of atoxyl was not as Bechaps assumed an anilide of arsenic acid, but p-amino-phenylarsonic acid.

In genral, when primary aromatic amines are fused with arsenic group takes up the para – position with respect to nitrogen with the formation of p-position is already occupied, then either no substitution occurs or the corresponding o-amino derivative is obtained.

$$NH_2C_2H_5 + AsO_4H_3 \rightarrow NH_2C_6H_4.AsO_3H_2 + H_2O$$

Many other aromatic compounds such as phenols and certain indoles behave in the same manner. This is an extract parallel to the production of sulphanilic acid by the action of heat on aniline sulphate and may therefore be described as arsenation and the reaction products as aresenilic acid; most amino aryl arsonic acids known to have been prepared in this way.

Among other synthesis of aromatic arsenic derivatives the treating diazo compounds with arsenious acid or its salts the diaxo-group is replaced by arsenic to give primary arsonic acids.

$$CH_3\,N{:}\,NCl + As(OH)_2 \rightarrow CH_3.AsO(OH)_2 + N_2 + HCl$$

This reaction may also be used to prepare the corresponding antimony derivatives. Acids containing both these metals are obtained by the interaction of antimony oxide with diazotized amino phenyl antimonic acids; in this manner phenylene-arsonic-antimonic acids are formed. ($C_6H_4.AsO\,SbO_2.nH_2O$).

Aromatic amino arsonic acids are very reactive and may be used in a variety of synthesis. In particular they are readily diazotized and the diazoaryl arsonic acids so obtained undergo the usual decomposition of diazo compounds and couple in the normal manner to give azo-dyes, all of which are soluble in alkalies owing to the presence of the arsonic acid residue H_3AsO_3. In its properties this group shows a general resembles to the sulphonic group SO_3H and is some reductions to the carboxyl group %COOH .

P-Aminophenylarsonic acid, Arsanilic acid $H_2N.C_6H_4AsO_3H_2$ is the most important of these acids, as ready described, is obtained by the arsenation of aniline.

It crystallizes in colourless needles and is sparingly soluble in cold water. But in hot water it dissolves easily.

Physical Properties

Molecular Formula	= $C_6H_8AsNO_3$
Formula Weight	= 217.05422
Composition	= C(33.20%) H(3.71%) As(34.52%) N(6.45%) O(22.11%)
Monoisotopic Mass	= 216.972015 Da
Nominal Mass	= 217 Da
Average Mass	= 217.0542 Da
Log P	=-1.28

Chemical Properties

In hot water it dissolves easily and produces an acid solution. As an acid the compound is readily soluble in alkalies but it also possesses basic properties, as shown by its solubility in an excess of dilute mineral acid. It does not melt at any particular temperature, but decomposes in the neighbourhood of 300°C. The sodium salt was formerly employed in medicine under the name **atoxyl** in cases of syphilis and sleeping sickness. In the year 1902, when this substance was first placed on the market, experiment were being carried out by Ehrlich and Siga with the object of curing parasitic diseases by the injection of suitable chemical compounds. In this work the action compounds. In this work the action of the atoxyl on trypanosomes was investigated, with results which led to further experiments with arsenic derivatives.

Reduction products of Arsanilic acids: *pp'* Diamino arsenobenzene

On energetic reduction the aryl arsenic acids are converted (Arsanilic acid) being *pp'* diamine arsenobenzene or *p*-arsenoaniline possessing the structure.

$$H_2N-C_6H_4-As{=}As-C_6H_4-NH_2$$

This compound can be prepared in a variety of ways, such as by the reduction of *p*–aminophenyl arsonic acid with sodium hydrosulphite or with stannous chloride and hydriodic acid. It melts at 260°C, and is also insoluble in water and aqueous alkali but as a base is readily, soluble in dilute hydrochloric acid. Oxidizing agents attack it rapidly, as the arseno compounds in general are characterized by strong reducing properties. Diamino arsenobenzene also gives the reactions of primary amines. It is readily diazotized, converted into azo dyes and condensed with aldehyde.

The reduction of *p*-amino phenyl arsonic acid to diamine arsenobenzene is found to bring about a great increase in toxic power and also in trypanocidal[1] action, in explanation of which it was been suggested that the chemoreceptors of the parasites are able to attach themselves to the trivalent but not to the pentavalent arsenic residue. The belief that only those radicals containing of other arsenic compounds and to the isolation of salvarsan.

p-Diahydrooxy *m*-diaminobenzene (salvarsan) (Arsphenamine, also known as Salvarsan and 606)

History

Sahachiro Hata discovered the anti-syphilitic activity of this compound in 1908 in the laboratory of Paul Ehrlich, during a survey of hundreds of newly-synthesized organic arsenical compounds. Ehrlich had theorized that by screening many compounds a drug could be discovered with anti-microbial activity. Ehrlich's team began their search for such a "magic bullet" among chemical derivatives of the dangerously toxic drug atoxyl. This was the first organized team effort to optimize the biological activity of a lead

[1]Causing blood transmitted diseases in human being caused by parasite poisons.

compound through systematic chemical modifications, the basis for nearly all modern pharmaceutical research.

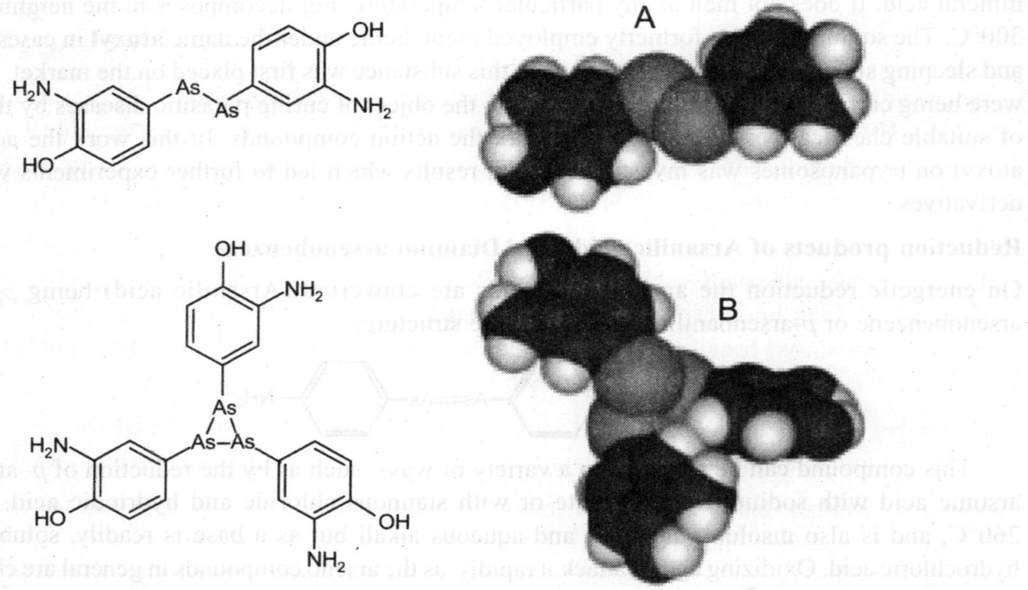

Arsphenamine was marketed under the trade name *Salvarsan* in 1910. It was also called *606*, because it was the 606th compound synthesized for testing. Salvarsan was the first organic anti-syphillitic, and a great improvement over the inorganic mercury compounds that had been used previously. A more soluble (but slightly less effective) arsenical compound, Neosalvarsan, (*neoarsphenamine*), became available in 1912. These arsenical compounds came with considerable risk of side effects, and they were supplanted as treatments for syphilis in the 1940s by penicillin.

The bacterium that causes syphilis is a spirochete, *Treponema pallidum*. Arsphenamine is not toxic to spirochetes until it has been converted to an active form by the body.

After leaving Erlich's laboratory, Hata continued parallel investigation of the new medicine in Japan.

Structure

The structure was believed to feature an As=As bond. However, in 2005, it was shown to be a mixture of the cyclic trimer and a pentamer. The revised structure features As-As single bonds, not double bonds.

Synthesis

This compound is prepared from the *p*-hydroxy-phenyl-arsenic acid. The later was obtained directly from phenol and arsenic acid in the same manner as phenol-sulphonic acid is obtained from phenol and sulphuric acid. It was then nitrated and resulting *p*-hydroxy-*m*-nitrophenyl arsonic acid reduced as indicated in the following equations:

phenol $\xrightarrow{+AsO(OH)_3}$ p-hydroxy phenol arsonic acid $\xrightarrow[\text{Conc } HNO_3]{\text{Conc } H_2SO_4}$ p-hydroxy-m-nitrophenyl arsonic acid $\xrightarrow[\text{or } Na_2S_2O_4]{\text{reduction with Na amalgam}}$ p-hydroxy aminoarsonic acid

p-hydroxy aminoarsonic acid ⇅ (reduction with $Na_2S_2O_4$ / oxidation with H_2O_2)

p-hydroxy-m-nitrophenyl arsonic acid → p-dihydrooxy-m-arsenobenzene (salvarson)

Organophosphorus Compounds

Organophosphorus compounds are chemical compounds containing carbon-phosphorus bonds. Organophosphorus chemistry is the corresponding science exploring the properties and reactivity of organophosphorus compounds. Phosphorus shares group 15 in the periodic table with nitrogen and phosphorus compounds and nitrogen compounds have much in common.

Phosphorus can adopt oxidation states –3, –1, 1, 3 and 5. In chemical literature very often compounds with +3 or –3 oxidation state are grouped together as having a (III) oxidation state regardless of sign. In an official and more descriptive nomenclature phosphorus compounds are identified by their coordination number δ and their valency λ. In this system a phosphine is a $\delta^3\lambda^3$ compound.

Phosphanes and Phosphines

The parent compound of the **phosphanes** is PH_3, called phosphine in the US and UK and phosphane elsewhere.[3] Replacement of one or more protons by an organic residue, gives $PH_{3-x}R_x$, an organophosphine or organophosphane, again depending on the country. The phosphorus atom in phosphanes/phosphines has a formal oxidation state –3 ($\delta^3\lambda^3$) and are the phosphorus analogues of simple amines.

An often used organic phosphine is triphenylphosphine. Like amines, phosphines have a trigonal pyramidal molecular geometry although with larger angles. The C-P-C bond angle is 98.6° for trimethylphosphine increasing to 109.7° when the methyl groups are replaced by *tert*-butyl groups. Commonly the size of these ligands are described by the parameter called cone angle.

The barrier to inversion is also much higher than in amines for a process like nitrogen inversion to occur and therefore phosphines with three different substituents can display optical isomerism.

The basicity of phospines is less than that of corresponding amines, for instance phosphonium ion itself has a pKa of -14 compared to 9.21 for ammonium ion; trimethylphosphonium has a pK_a of 8.65 compared to that of 9.76 of trimethylamine. However, triphenylphosphine (pK_a 11.2) is more basic than triphenylamine (pK_a ?), mainly because the lone pair of the nitrogen in NPh_3 is delocalized throughout the three phenyl rings.

Amines and phosphines both have a lone pair of electrons but with a difference. Whereas the lone pair in an amine shares its electrons at every opportunity for delocalization for instance in pyrrole, the phosphorus atom in a similar configuration will not. For this reason, the phosphorus equivalent of pyrrole called phosphole is not at all aromatic.

The reactivity of phosphines match that of amines with regard to nucleophilicity in the formation of Phosphonium salts with the general structure $PR_4^+X^-$. This property is used in the Appel reaction converting alcohols to alkyl halides.

A difference in reactivity with amines is the ease of oxidation of phosphines to phosphine oxides.

Synthesis

Synthetic produces for phosphines are:

- Nucleophilic displacement of phosphorus halides with organometallic reagents such as Grignard reagents.
- Nucleophilic displacement of metal phosphides, generated by reaction of potassium metal with phosphine as in sodium amide with alkyl halides.

- Nucleophilic addition of phosphine with alkenes in presence of a strong base (often KOH in DMSO), Markovnikov's rules apply.[4] Phosphine can be prepared in situ from red phosphorus and potassium hydroxide. Primary (RPH_2) and secondary phospines (RRPH) do not require a base with electron-deficient alkenes such as acrylonitriles.

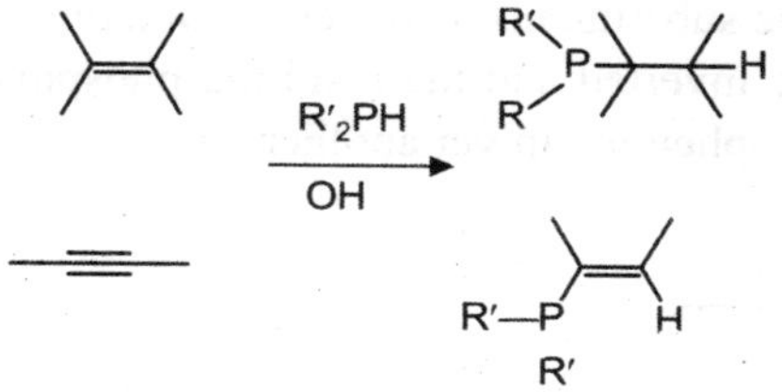
R'₂PH
OH
R'
P
R'
H
R'—P
H
R'

- Radical addition of phosphines to alkenes with AIBN or organic peroxides to give anti-Markovnikov adducts.
- Nucleophilic addition of phosphine and phosphines to alkynes in presence of base. Secondary phosphines react with electron-deficient alkynes such as phenylcyanoacetylene without base.
- Organic reduction of phosphine oxides for instance with chlorosilane.

Reactions

The main reaction types of phosphines are:

- as nucleophiles for instance with alkyl halides to phosphonium salts. Phospines are key nucleophilic catalysts in the dimerization of enones in the Rauhut-Currier reaction.
- as reducing agents:

Phosphines are reducing agents in the Staudinger reduction converting azides to amines and in the Mitsunobu reaction converting alcohols into esters. In these processes the phosphine is oxidized to the

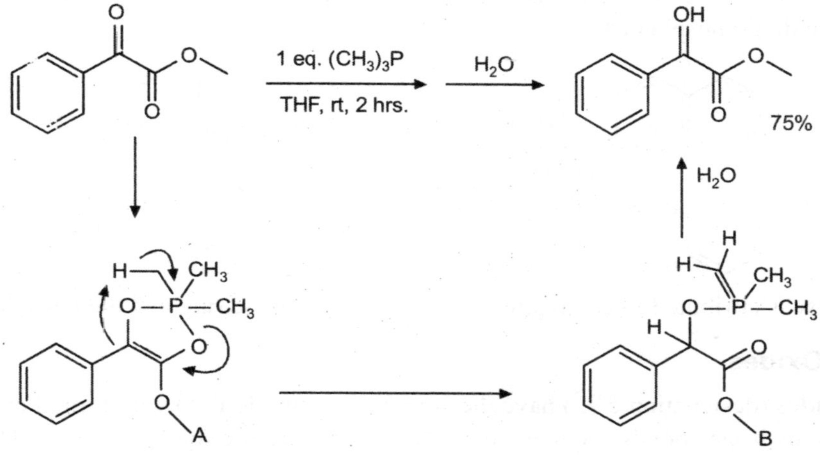
1 eq. (CH3)3P
THF, rt, 2 hrs.
H2O
OH
75%
H2O
CH3
O—P—CH3
A
H
CH3
O—P—CH3
B

phosphine oxide. Phosphines have also been found to reduce activated carbonyl groups for instance the reduction of an α-keto ester to an α-hydroxy ester in *scheme 2*. In the proposed reaction mechanism the first proton is on loan from the methyl group in trimethylphosphine (triphenylphosphine does not react).

When modified with suitable substituents as in certain *diazaphospholenes* (*scheme 3*) the polarity of the P-H bond can actually be inverted and the resulting phosphine hydride can reduce a carbonyl group as in the example of benzophenone in yet another way.

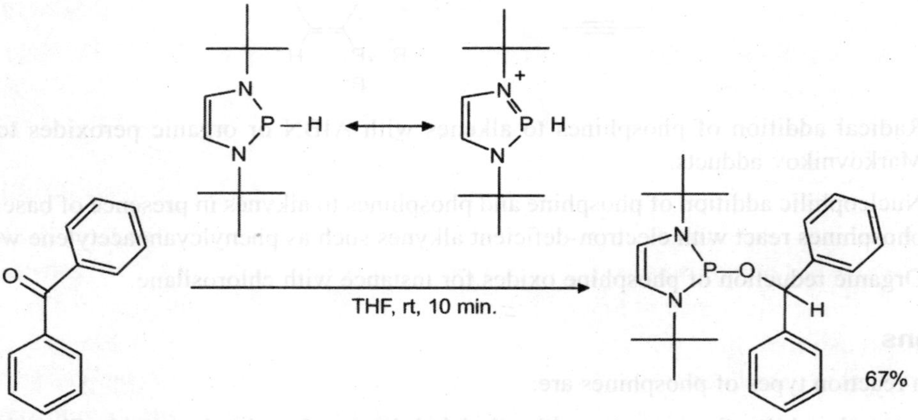

THF, rt, 10 min.

67%

- Multidentate phosphines such as diphosphines and BINAP are important ligands in organometallic chemistry.

Phosphanes

Primary phosphanes are under-used in chemistry due to their general lack of stability towards oxygen. One study reports on several novel air-stable aromatic primary phosphanes prepared by organic reduction of the corresponding phosphonate:

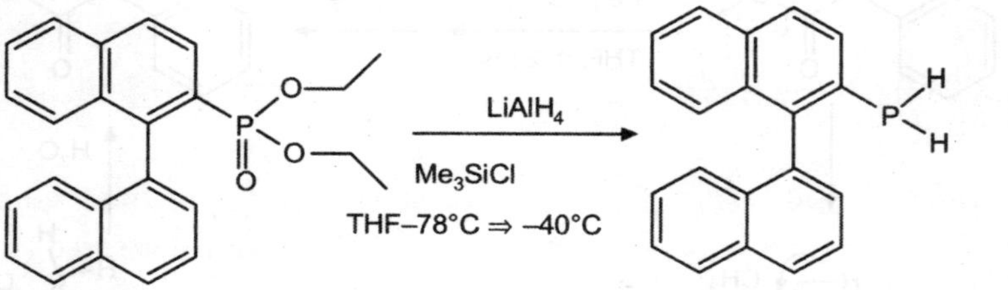

$LiAlH_4$

Me_3SiCl

THF–78°C ⇒ –40°C

The stability is attributed to conjugation between the aromatic ring and the phosphorus lone pair.

Phosphine Oxides

Phosphine oxides (designation $\delta^3\lambda^3$) have the general structure $R_3P{=}O$ with formal oxidation state 1. Phospines form hydrogen bonds and many phosphines are therefore soluble in water. The P=O bond is very polar with a dipole moment of 4.51 D for triphenylphosphine.

The nature of the phosphorus to oxygen double bond is a matter of debate. Pentavalent phosphorus like nitrogen is not compatible with the octet rule. In older literature the bond is represented as a dative bond just like an amine oxide. The prevailing view is that of a full double bond with back bonding taking place between a filled oxygen electron pair and an empty phosphorus d-orbital (lacking in nitrogen). problem is: the P=O bond does not react as any C=C double bond as addition reactions are absent and involvement of a phosphorus d-orbital in bonding is not supported in silico. Alternative theories favor an ionic bond $P^+–O^-$ which on its own strength should explain the short bond length. Molecular Orbital Theory proposes that the short bond length is attributed to the donation of the lone pair electrons from oxygen to the antibonding phosphorus-carbon bonds. This proposal is supported by *ab initio* calculations and has gained consensus in the chemistry community.

Phospines are easily oxidized to phosphine oxides as examplified by the directed synthesis of a phospha crown, the phosphorus analogue of an aza crown where it is not possible to isolate the phosphine itself.

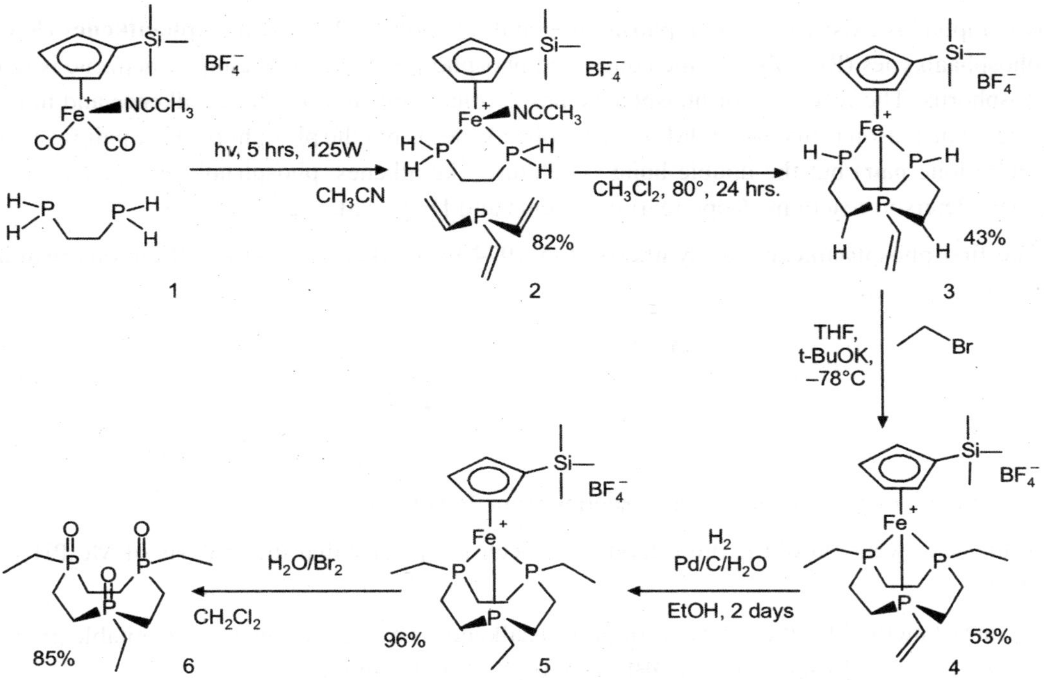

Phosphonates

Phosphonates have the general structure $R–P(=O)(OR)_2$. In the Horner-Wadsworth-Emmons reaction and the Seyferth-Gilbert homologation phosphonates are used as stabilized carbanions in reactions with carbonyl compounds. Phosphonates have many technical applications and bisphosphonates are a class of drugs.

Phosphite and Phosphate Esters

Phosphite esters or phosphites have the general structure $P(OR)_3$ with oxidation state +3. Phosphites are employed in the Perkow reaction and the Arbusov reaction. Phosphate esters with the general structure $P(=O)(OR)_3$ and oxidation state +5 are of great technological importance as flame retardant agents and plasticizers. Lacking a P–C bond, these compounds are technically not organophosphorus compounds.

Phosphoranes

Phosphoranes have a –5 oxidation state ($\delta^5\lambda^5$) with parent compound the non-stable **phosphoran** PH_5 or **λ^5-phosphane** (lambda 5 phosphane). Phosphorus ylides are unsaturated phosphoranes used in the Wittig reaction.

Phosphorus Multiple Bonds

Many compounds exist with carbon phosphorus multiple bonds (P=C) as **phosphaalkenes** (R_2C::PR) and phosphaalkynes (RC:::PR). In the compound phosphorine one carbon atom in benzene is replaced by phosphorus. The reactivity of phosphaalkenes is often compared to that of alkenes and not to that of imines. The reason is that the HOMO of phosphaalkenes is not the phosphorus lone pair (as in imines the amine lone pair) but the double bond. Therefore like alkenes, phosphaalkenes engage in Wittig reactions, Peterson reactions, Cope rearrangements and Diels-Alder reactions.

The first phosphaalkene was synthesised in 1974 by Becker as a keto-enol tautomerism akin a Brook rearrangement:

O tms
R—P=
tms O
P—
R

with R = methyl or phenyl and tms representing trimethylsilyl.

In the same year Harold Kroto established spectroscopically that thermolysis of Me_2PH yielded CH_2=PMe.

A general method for the synthesis of phosphaalkenes is by 1,2-elimination of suitable precursors, initiated thermally or by base such as DBU, DABCO or triethylamine:

mes Ph
P—C—H
Cl Ph
DBU
–DBU.HCl
mes Ph
P=C
Ph

The Becker method is used in the synthesis of the phosphorus pendant of Poly (p-phenylene vinylene).

85°C, 2 hrs.

Diphosphenes are compounds containing P=P phosphorus double bonds. Phosphazenes have a P=N double bond.

Phosphine Donors

empty d orbitals on phosphine can act as π-acceptor orbitls } not very important unless R-groups are electron-withdrawing

Phosphine ligands
Excellent soft-donor ligands with a wide variety of easily adjusted electronic factors

Neutral $2e^-$ donor

R=carbon gropus { phosphine (US) / phosphane (Germany/Europe)

R=OR groups ⟶ phosphite

Because of the three R-groups on the phosphine ligand and the overall tetrahedral coordination geometry it is the most versatile of the neutral 2-electron donor ligands. Variation of the three R-groups can effect: large changes in the donor/acceptor properties of the phosphine (from excellent donor/poor π-acceptor to poor donor/excellent p-acceptor) large changes in the steric profile of the phosphine (from fairly small to enormous) generation of a large number of polydentate polyphosphines (bis-, tris-, tetra-, penta-, and hexaphosphine ligands are all known) that can adopt specific coordination geometries (cis-enforcing, facial tridentate, bridging, bridging and chelating, etc.)

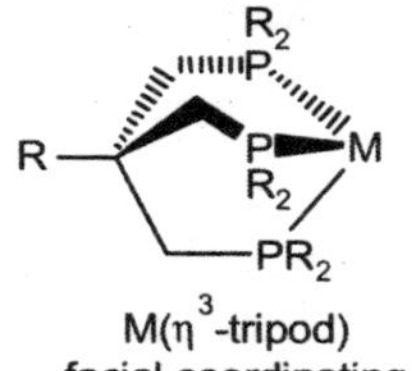

M(η^3-tripod)
facial coordinating

Racemic-M_2(P4)
binucleating phosphine
able to bridge and chelate 2 metals

Tolman's Cone Angle and Electronic Parameter

In 1977 Chad Tolman (Dupont Chemicals) published a classic review article covering methods that he developed for ordering a wide variety of phosphine ligands in terms of their electron-donating ability and steric bulk.

The electron-donating ability of a phosphine ligand was determined by reacting one equivalent of the phosphine (mono-dentate only) with $Ni(CO)_4$ to make a $Ni(CO)_3$(phosphine) complex. He then measured the carbonyl nCO IR stretching frequency (the very sharp a_1 high energy mode) of the $Ni(CO)_3$(phosphine) complex. The more electron density the phosphine ligand donated to the metal center, the more π-back-bonding occurred to the carbonyl ligands, weakening the C≡O triple bond, thus lowering the ν_{CO} IR stretching frequency.

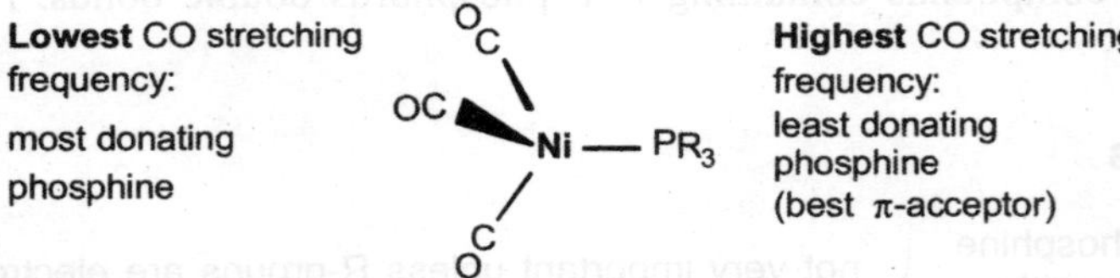

The size or steric bulk of a phosphine ligand was determined from simple 3-D space-filling models of the phosphine ligands. Tolman coined the name cone angle (θ) to indicate the approximate amount of "space" that the ligand consumed about the metal centre.

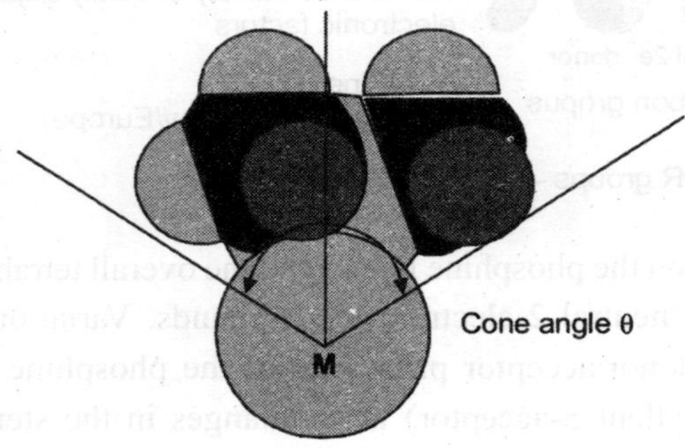

Phosphine Donor Ability Ranked by Tolman's Electronic Parameter θ (most donating to least)

PR_3	mixed	$P(OR)_3$	PX_3	θ, cm^1
$P(t\text{-}Bu)_3$				2056.1
PCy_3				2056.4
$P(o\text{-}OMe\text{-}C_6H_4)_3$				2058.3
$P(i\text{-}Pr)_3$				2059.2
PBu_3				2060.3
PEt_3				2061.7
	PEt_2Ph			2063.7

PR_3	mixed	$P(OR)_3$	PX_3	θ, cm^1
PMe3				2064.1
	PMe_2Ph			2065.3
$P(p\text{-}OMe\text{-}C_6H_4)_3$	$PPh_2(o\text{-}OMe\text{-}C_6H_4)$			2066.1
PBz_3				2066.4
$P(o\text{-}Tol)_3$				2066.6
$P(p\text{-}Tol)_3$	$PEtPh_2$			2066.7
	$PMePh_2$			2067.0
$P(m\text{-}Tol)_3$				2067.2
	$PPh_2(NMe_2)$			2067.3
	$PPh_2(2,4,6\text{-}Me\text{-}C_6H_2)$			2067.4
	$PPhBz_2$			2067.6
	$PPh_2(p\text{-}OMe\text{-}C_6H_4)$			2068.2
	PPh_2Bz			2068.4
PPh_3				2068.9
	$PPh_2(CH{=}CH_2)$			2069.3
$P(CH{=}CH_2)_3$	$PPh_2(p\text{-}F\text{-}C_6H_4)$			2069.5
	$PPh(p\text{-}F\text{-}C_6H_4)_2$			2070.0
$P(p\text{-}F\text{-}C_6H_4)_3$				2071.3
	$PPh_2(OEt)$			2071.6
	$PPh_2(OMe)$			2072.0
	$PPh(O\text{-}i\text{-}Pr)_2$			2072.2
$P(p\text{-}Cl\text{-}C_6H_4)_3$				2072.8
	PPh_2H			2073.3
	$PPh(OBu)_2$			2073.4
$P(m\text{-}F\text{-}C_6H_4)_3$				2074.1
	$PPh(OEt)_2$			2074.2
	$PPh_2(OPh)$			2074.6
	$PPh_2(C_6F_5)$			2074.8
		$P(O\text{-}i\text{-}Pr)_3$		2075.9
		$P(OEt)_3$		2076.3
	$PPhH_2$			2077.0
$P(CH_2CH_2CN)_3$				2077.9
		$P(OCH_2CH_2OMe)_3$		2079.3
		$P(OMe)_3$		2079.5

PR_3	mixed	$P(OR)_3$	PX_3	θ, cm^1
	$PPh(OPh)_2$			2079.8
		PPh_2Cl		2080.7
		PMe_2CF_3		2080.9
		$P(O\text{-}2,4\text{-}Me\text{-}C_6H_3)_3$	PH_3	2083.2
		$P(OCH_2CH_2Cl)_3$		2084.0
		$P(O\text{-}Tol)_3$		2084.1
		$P(OPh)_3$		2085.3
		$P(OCH_2)_3CR$		2086.8
		$P(OCH_2CH_2CN)_3$		2087.6
	$P(C_6F_5)_3$			2090.9
			PCl_3	2097.0
			PF_3	2110.8

Problem: Order the following phosphines from strongest to weakest ? donor:

$P(OEt)_3$ PPh_3 PPr_3 PCl_3 $PPh(OMe)_2$

Phosphine Steric Bulk Ranked by Tolman's Cone Angle θ (smallest to largest)

PR_3	mixed	$P(OR)_3$	PX_3	(°)
			PH_3	87
	$PPhH_2$	$P(OCH_2)_3CR$		101
			PF_3	104
	$Me_2PCH_2CH_2PMe_2$	$P(OMe)_3$		107
		$P(OEt)_3$		109
	$P(CH_2O)_3CR$			114
	$Et_2PCH_2CH_2PEt_2$			115
	$P(OMe)_2Ph$ or Et			115
	$PPh(OEt)_2$			116
PMe3				118
	$Ph_2PCH_2PPh_2$			121
	PMe_2Ph			122
		PMe_2CF_3	PCl_3	124
	$Ph_2PCH_2CH_2PPh_2$			125
	PPh_2H	$P(OPh)_3$		128

PR_3	mixed	$P(OR)_3$	PX_3	(°)
		$P(O\text{-}i\text{-}Pr)_3$		130
			PBr_3	131
PEt_3, PPr_3, PBu_3	$PPh_2(OMe)$			132
	$PPh_2(OEt)$			133
	PEt_2Ph, $PMePh_2$			136
$P(CF_3)_3$				137
	$PEtPh_2$			140
	$Cy_2PCH_2CH_2PCy_2$			142
PPh_3				145
	$PPh_2(i\text{-}Pr)$			150
	$PPh_2(t\text{-}Bu)$			157
	$PPh_2(C_6F_5)$			158
$P(i\text{-}Pr)_3$				160
PBz_3				165
PCy_3	$PPh(t\text{-}Bu)_2$			170
		$P(O\text{-}t\text{-}Bu)_3$		175
$P(t\text{-}Bu)_3$				182
	$P(C_6F_5)_3$			184
$P(o\text{-}Tol)_3$				194
$P(mesityl)_3$				212

Commonly Used *Monodentate* Phosphines:

PPh_3 (145°, medium donor), triphenylphosphine, tpp "The KING"

air-stable, white crystalline material, no odor to speak of

Increasing σ-Donor Ability:

$PMePh_2$ (136°), PMe_2Ph (122°), PMe_3 (118°), PEt_3 (132°)

$P(Cy)_3$ (170°) tricyclohexylphosphine, $P(t\text{-}Bu)_3$ (182°)

the alkyl phosphines are strong σ-donors; usually colourless liquids, somewhat to very air-sensitive, horrible smelling (unless very high MW and non-volatile)

Poor σ-Donors, Good π-Acceptors:

Phosphites: $P(OMe)_3$ (107°), $P(OEt)_3$ (110°), $P(OPh)_3$ (128°)

phosphites are relatively poor σ-donors, but can be fairly good π-acceptor ligands (about half as good as CO); low MW ones are usually colorless liquids, higher MW compounds are white solids; usually air-stable but moisture sensitive; sometimes sweet smelling

PF3 (104°) } v. poor donor; strong π-acceptor, almost as good as CO

Commonly Used Polydentate Phosphines:

Ph2P
PPh2

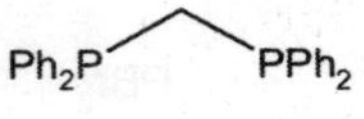

dppm (121°)
diphenylphosphinomethane
bis(diphenyl)phosphinomethane
bridging ligand

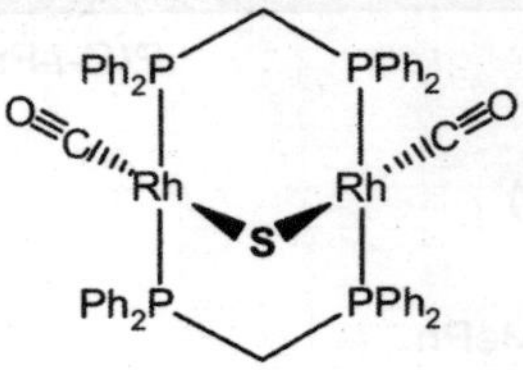

Ph2P
PPh2
O≡C
C≡O
Rh
Rh
S
Ph2P
PPh2

A-*Frame bimetallic*

$Rh_2(\mu\text{-}S)(CO)_2(dppm)_2$

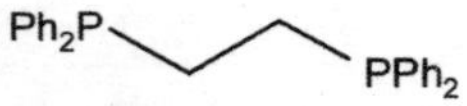

Ph2P
PPh2

dppe (125°)
diphenylphosphinoethane
bis(diphenyl)phosphinoethane
chelating ligand

86.9°
Ph2P
PPh2
Ni
Cl
Cl
95.5°

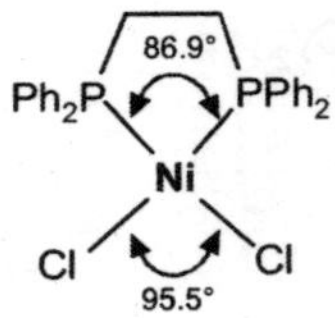

Typical P-M-P angle for a 5-membered chelate ring 82–87°

$NiCl_2$(dppe)

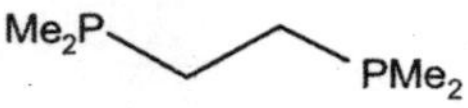

Me2P
PMe2

dmpe (107°)
dimethylphosphinoethane
bis(dimethyl)phosphinoethane
chelating ligand
electron-rich, strong donor

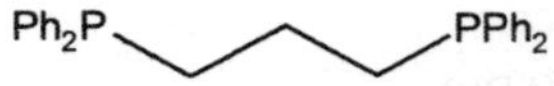

Ph2P
PPh2

dppp (127°)
diphenylphosphinopropane
bis(diphenyl)phosphinopropane
chelating ligand
forms 6-membered rings

typical P-M-P angle for a 6-membered chelate ring 88–92°

Some Other Polydentate Phosphines:

Facial coordinating mode

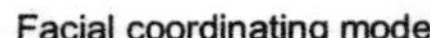

Te
Te
Ni
PPh
Ph2P
Ph2P

Ph
P
Ph2P
PPh2

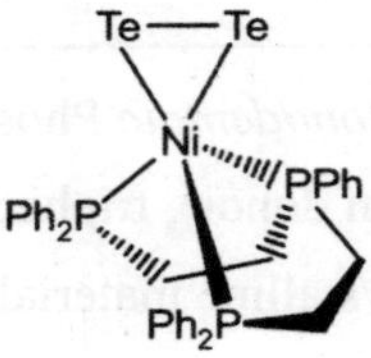

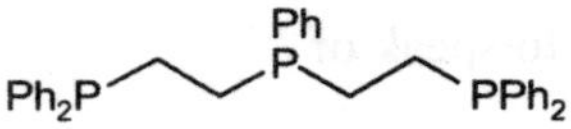

triphos
bis(diphenylphosphinoethy) phenylphosphine
***bis*-chelating ligand**

Ph2P
+
Cl
Ni
P
Ph
Ph2P

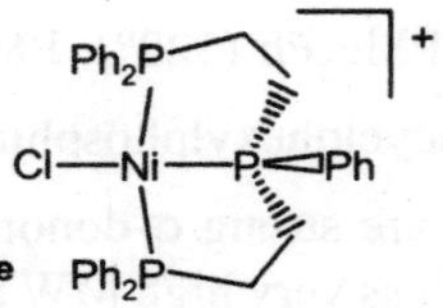

Bertinesson
planar coordinating mode

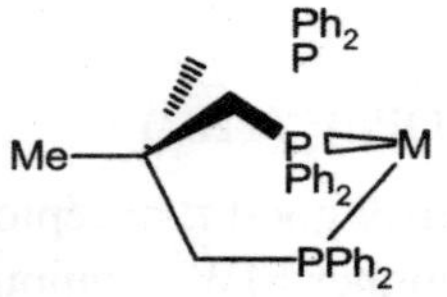

Ph2
P
Me
P
M
Ph2
PPh2

tripod
tris(diphenylphosphinoethyl)methane
***bis*-chelating ligand**
facial coordination

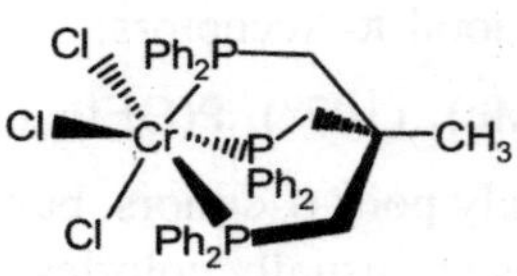

Cl
Ph2P
Cl
Cr
P
CH3
Ph2
Cl
Ph2P

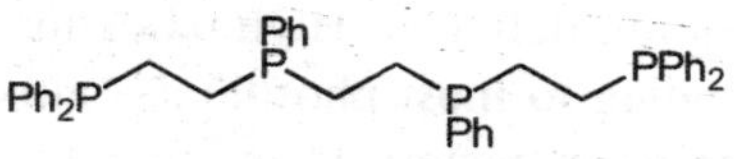

tetraphos-1
1,1,4,7,10,10-hexaphenyl-1,4,7,10-tetraphophadecane
tris-chelating or binucleating (bridging) ligand

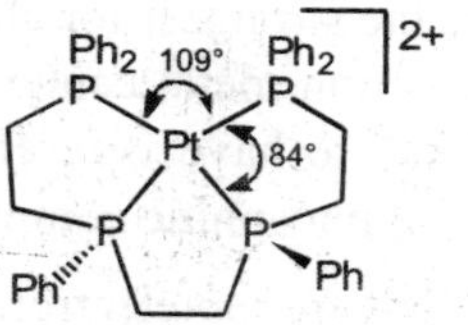

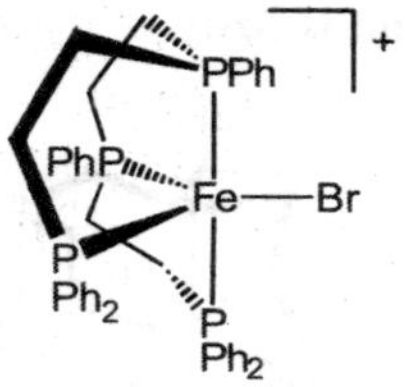

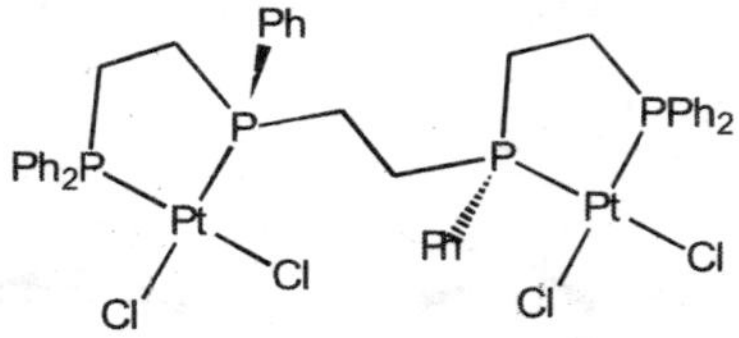

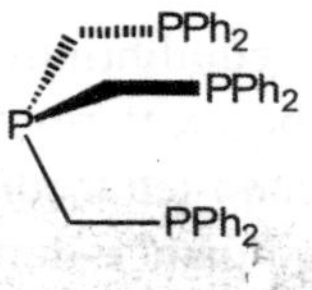

Tetraphos-2
tris(diphenylphosphinoethyl)phosphine
tris-chelating ligand
facial coordination

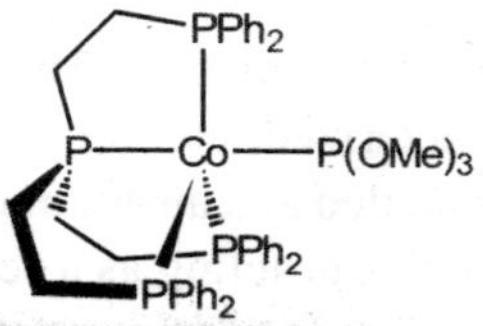

Some Structural Issues

Phosphines have only been characterized as simple 2 e^- donating, terminal-only ligands. No true μ_x-bridging monophosphines are known (although bridging ***phosphides***, PR_2^-, are very common).

Phosphines generally tend to orient *trans* to one another in order to minimize steric interactions (especially true for bulky PR_3). Chelating bisphosphine ligands are used to enforce *cisoidal* coordination geometries when needed.

Some typical first row M-PR_3 average bond distances:

Ti-P	2.6 Å
V-P	2.5 Å
Cr-P	2.4 Å
Ni-P	2.1 Å

M-P distances decrease due to the contraction of the metal atom radius as one proceeds to the right and the atoms become more electronegative. Distances also decrease due to stronger M-P bonding as one moves to the right across the transition metal series (later transition metals are "softer" and prefer bonding to phosphines).

M-P bonds are the strongest for alkylated phosphine ligands bonding to a neutral or monocationic middle to later transition metal center that is electron-deficient. High oxidation state early transition metals are too "hard" to have very effective bonding to most phosphines, although more and more early transition metal phosphine complexes are being characterized and found to be reasonably stable.

Metal centers that are *too* electron-rich will generally not want to have a strong electron-donating alkylated phosphine coordinated, this leads to weaker M-P bonding and more likely phosphine dissociation. One example is $Rh_2(\mu\text{-}CO)(CO)_4(P_4)$ shown below, that rearranges to form the asymmetric μ,η^3,η^1–P_4 coordinated dimer, shown to the right, when one CO ligand is lost:

This is a fully reversible CO and temperature dependent equilibrium. The asymmetric dimer, which has been structurally characterized, can be considered to be zwitterionic: one cationic Rh(+1) center that has 3 phosphines coordinated and an anionic Rh(−1) pseudo-tetrahedral center that has 3 π-accepting CO ligands. One of these CO's is acting as a semi-bridging ligand *p*-donating some electron density to the other formally cationic square-planar rhodium.

Bond Length vs. Bond Strength

Aroney and coworkers (*Inorg. Chem.*, **1995**, *34*, 330-336) claimed that the shorter Cr-PCl_3 bond in $Cr(CO)_5(PCl_3)$ vs. $Cr(CO)_5(PMe_3)$ was a result of strong π-backbonding. They did NOT make any statements about bond strengths, but implied that the Cr-PCl_3 bond is shorter and, therefore, stronger than the Cr-PMe_3 bond.

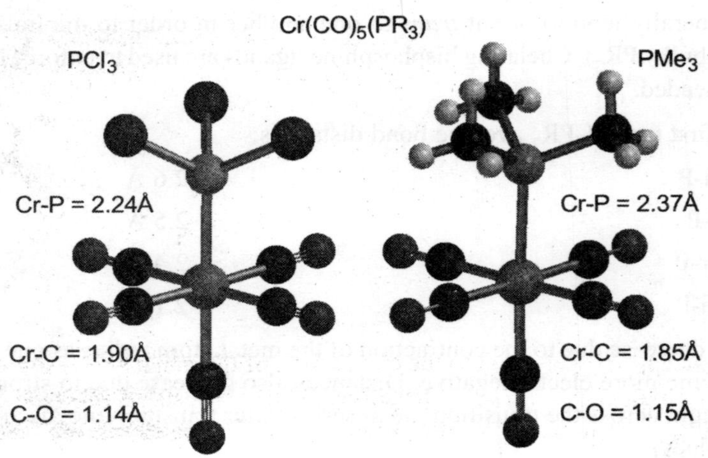

Let's compare the P lone pair orbital for free PCl_3 and PMe_3 ligand, which is the highest occupied molecular orbital (HOMO):

HOMO energy = ⁻8.36 eV ⁻6.23 eV

Note the higher energy (better π donor) and greater orbital extent of the P lone pair on PMe_3 (right plot). This means that the PMe_3 ligand does not have to get as close to the metal center as PCl_3 in order to get good bonding overlap. Below are shown single-point (no geometry optimization) DFT calculations on $Cr(CO)_5(PCl_3)$ and $Cr(CO)_5(PMe_3)$ based on their crystal structures, followed by calculations with the phosphine ligands dissociated (4.0 Å separation) to see the relative change in energies.

Note the clear-cut π-backbonding between the PCl_3 and the Cr centre, while there is just a trace of p-backbonding in the Cr-PMe_3 bond. Note that the Cr-CO ligand trans to the PCl_3 has a bit more orbital extent indicating a little more p-backbonding ability.

The difference in total energies represents the difference in PMe_3 vs. PCl_3 bond energies:

ΔΔE Bond dissociation = (ΔE(bonded-unbonded)PMe_3) – (ΔE(bonded-unbonded)PCl_3)

ΔΔE Bond dissociation = 6.8 Kcals (PMe_3 system having stronger bonding)

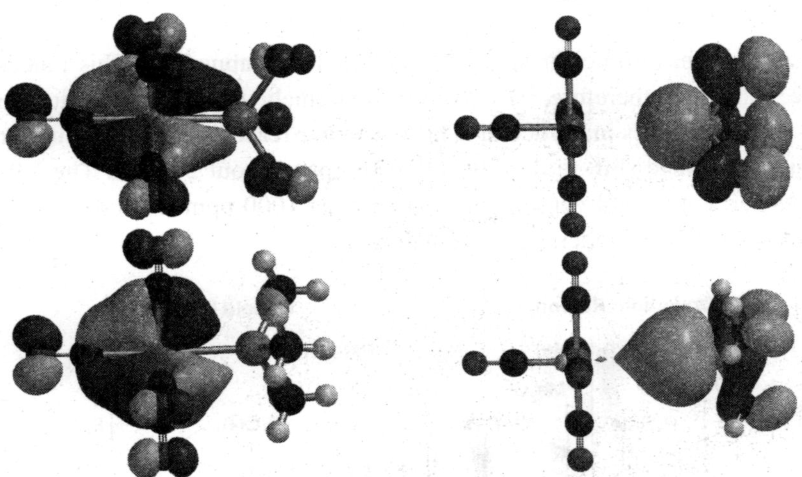

Further Test: Examine PCl_3 vs. PMe_3 bonding to a d^0 metal center: $TiCl_4(PCl_3)$ and $TiCl_4(PMe_3)$.

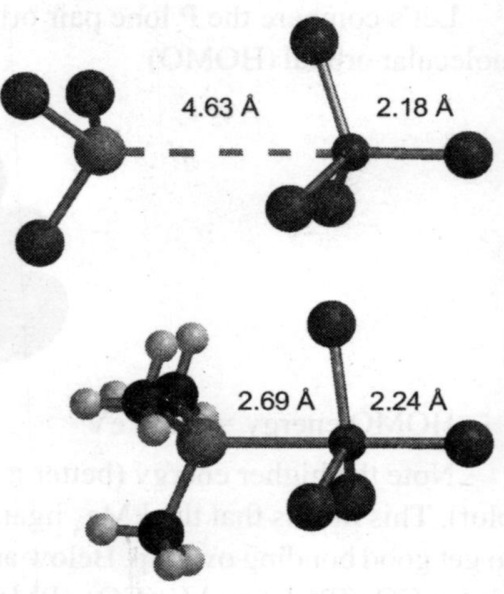

In this d^0 system the lack of any π-backbonding possibilities and good Ti-Cl bonding favoured complete dissociation of the poor p donating PCl_3 ligand. The far better PMe_3 π-donor, on the other hand, bonds nicely to the Ti(+4) d^0 centre.

Bond Length/Strength conclusions: For most systems a shorter bond *usually* indicates a stronger bond when comparing similar atoms and bonds. For metal-ligand complexes there can be exceptions to this when the ligands in question have fairly different donor/acceptor properties. In $Cr(CO)_5(PCl_3)$ the shorter bond distance relative to $Cr(CO)_5(PMe_3)$ arises due to the combination of a contracted lower energy P orbitals and moderate to significant π-backbonding. The DFT calculations indicate that the PMe_3 complex has stronger M-P bonding despite the significantly longer Cr-P distance (2.37Å vs. 2.24Å). The greater PMe_3 lone pair orbital extent and higher energy (better donor ability) produces a strong M-P π-bond with a longer bond distance relative to the PCl_3.

In order to get effective π-backbonding in the Cr-PCl_3 complex, the bond distance has to be shorter to get optimum orbital overlap. In the Ti(+4) d^0 system the PCl_3 ligand did not form a Ti-P bond, even though $TiCl_4$ only has an 8e- count, due to the lack of π-backbonding and weaker π-bonding ability relative to PMe_3.

31P NMR Spectroscopy

The ^{31}P (P-31) nucleus has a nuclear spin of ½ and is 100% abundant! This makes it functionally equivalent to the ^1H nucleus: therefore, ^{31}P NMR is an extremely valuable tool in studying phosphines in general, and M-PR_3 complexes in particular. This is another reason why phosphines are such valuable ligands.The typical chemical shift region for ^1H NMR spans about 20 ppm. The ^{31}P NMR chemical shift region, however, is much larger and can span almost 1000 ppm (including *phosphide* ligands)! Chemical shifts for some phosphines are listed below:

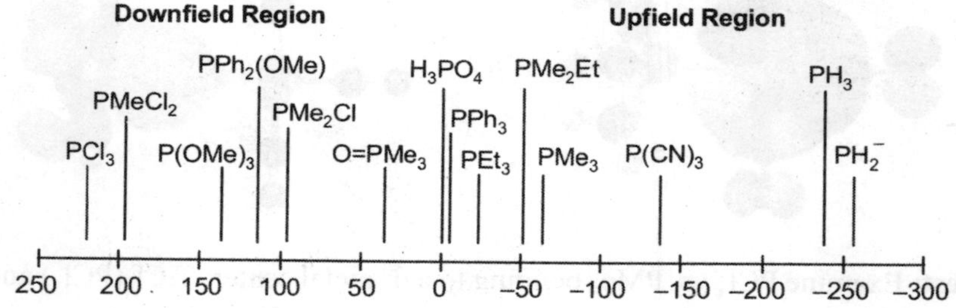

Ligand	Chemical Shift (ppm)	Ligand	Chemical Shift (ppm)
PCl_3	220	H_3PO_4	0 (reference)
$PMeCl_2$	191	$P(CF_3)_3$	-2
$PCy(OBu)_2$	184	PPh_3	-6
$P(OMe)_3$	140	PEt_3	-20
$P(OPh)_3$	126	$NaPPh_2$	-24
PEt_2Cl	119	$PMePh_2$	-28
$PPh_2(OMe)$	115	PPr_3	-33
PF_3	97	PMe_2Et	-50
PMe_2Cl	96	PMe_3	-62
$PMe_2(O\text{-}t\text{-}Bu)$	91	$P(CN)_3$	-135
$O{=}P(CH_2OH)_3$	45	PH_3	-238
$O{=}PMe_3$	36	KPH_2	-255

It is important to note that the positive and negative convention for chemical shifts for 31P NMR compounds changed in the early 1970's. In these older references, free PMe_3 would be listed with a +62 ppm chemical shift.

Chelate Ring Effects

The presence of chelate rings can have a significant effect on the ^{31}P NMR chemical shift position of M-P complexes.

4 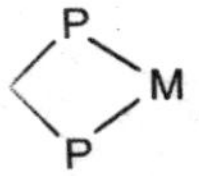 P–M–P upfield shift ~ –50 ppm

5 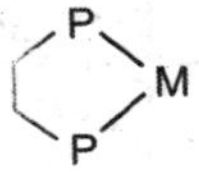 P–M–P downfield shift ~ +30 ppm

6 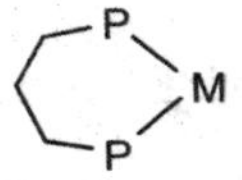 P–M–P upfield shift ~ –14 ppm

Similarly, the metal centre can also have a considerable shifting effect. Consider the following dppm chelates with 4-member rings:

$W(CO)_4(\eta^2\text{-dppm})$	23.7 ppm
$Mo(CO)_4(\eta^2\text{-dppm})$	0 ppm
$Cr(CO)_4(\eta^2\text{-dppm})$	23.5 ppm

^{31}P NMR, as with ^{1}H and ^{13}C methods, can be exceptionally useful in characterizing metal-phosphine complexes and structures:

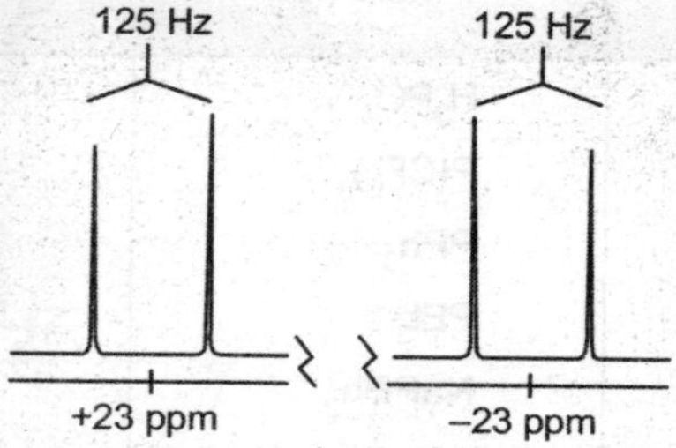

125 Hz 125 Hz

+23 ppm –23 ppm

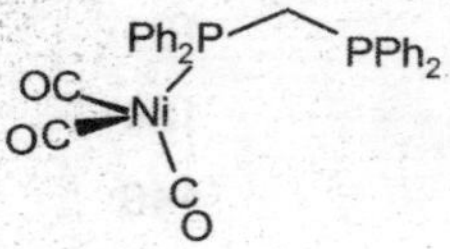

Ph_2P PPh_2 OC Ni OC C O

Organo Copper Silver and Gold Compounds

Copper, Silver, Gold. Univalent copper, silver, and gold form compounds of the type RM, and trivalent gold forms compounds of the types $RAuX_2$, R_2AuX, and R_3Au. These compounds are all highly unstable, the most stable of the gold compounds being the R_2AuX type.

It was suggested by Gibson and co-workers that trivalent gold compounds are all probably tetracovalent as the result of either dimerization or coordination with the solvent. Thus diethylgold bromide was assigned the structure:

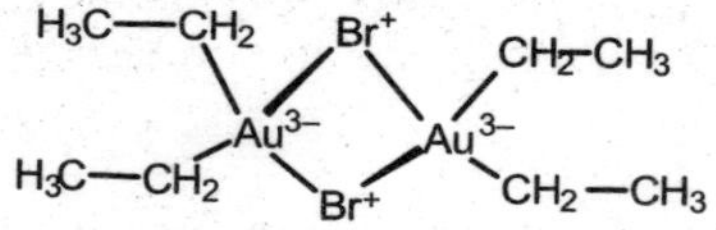

$H_3C—CH_2$ Br^+ $CH_2—CH_3$ Au^{3-} Au^{3-} $H_3C—CH_2$ Br^+ $CH_2—CH_3$

In keeping with this postulate is the observation by Gilman and Toods that trimethylgold can be stabilized by ethylenediamine, 2-aminopyridine, or benzylamineo The resulting complexes probably ha,oe structures similar to the following:

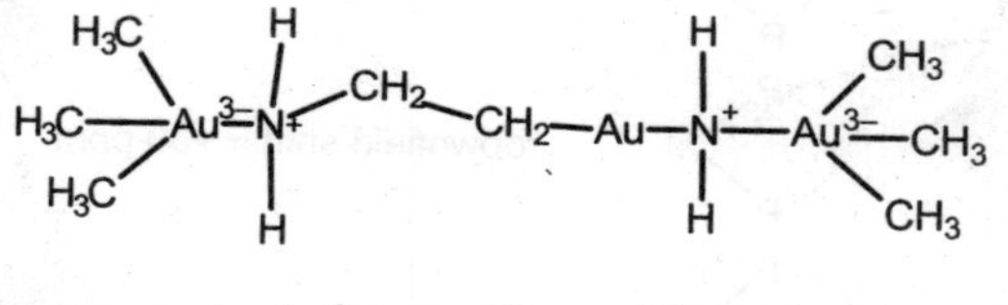

H_3C H H CH_3 $H_3C—Au^{3-}—N^+—CH_2—CH_2—Au—N^+—Au^{3-}—CH_3$ H_3C H H CH_3

Trimethylgold, which decomposes at about –40 to –35°C., is more stable than triethylgold.

Reactions. The most characteristic reaction of the univalent copper, silver, and gold compounds is their prompt and complete thermal decomposition to R–R compounds and the metal:

$$2C_6H_5M \rightarrow C_6H_5 - C_6H_5 + 2M$$

All of the organometallic compounds of this group are cleaved by acidic reagents at relatively low temperatures:

$$(CH_3)_3Au + HCl \xrightarrow{-65°C.} (CH_3)_2AuCl + CH_4$$

$$(CH_3)_3Au + C_6H_5SH \rightarrow (CH_3)_2AuSC_6H_5 + CH_4$$

As would be expected from the ionization potentials of the metals, the relative reactivities of organometallic compounds of Group IB metals are in the order Cu >Ag > Au.

This is illustrated in part by the following reaction equations:

$$\begin{aligned} & \quad -20^\circ C. \\ CH_3Cu + C_6H_5COCl &\rightarrow C_6H_5COCH_3 \\ C_6H_5Ag + C_6H_5COCl &\rightarrow (C_6H_5)CO \\ (CH_3)aAu + C_6H_5COCl &\rightarrow \text{no reaction} \\ C_6H_5Cu + CH_2{=}CHCH_2Br &\rightarrow CH_2{=}CHCH_2C_6H_5 \end{aligned}$$

Preparation. The organometallic derivatives of copper, silver, and gold are best prepared from the Grignard reagent or organolithium compounds:

$$\begin{aligned} C_6H_5MgI + CuI &\rightarrow C_6H_5Cu + MgI_2 \\ C_6H_5MgBr + AgBr &\rightarrow C_6H_5Ag + MgBr_2 \\ 2RLi + AuBr_3 &\rightarrow R_2AuBr + 2\,LiBr \\ (CH_3)_3AuBr + CH_3Li &\rightarrow (CH_3)_3Au + LiBr \end{aligned}$$

Organo Mercuric Compounds

Mercury

In 1853 Frankland reported the first unequivocal preparation of an organomercury compound. He prepared methylmercuric iodide by treating methyl iodide with mercury in sunlight:

$$CH_3I + Hg \rightarrow CH_3HgI$$

In 1854 Strecker made ethylmercuric iodide by treating ethyl iodide with metalhc mercury in diffused light. In the same year Dunhaupt obtained ethylmercuric chloride by heating triethylbismuth with mercuric chloride. Mercury chemistry then rapidly expanded to such an extent that Whitmore in his monograph on mercury compounds in 1921 stated that it was not possible for one man to cover the field entirely. Actually the existing literature on mercury is probably double in bulk that reported so ably by Whitmore. A part of the stimulus for research in this field arose from the discovery that certain mercury compounds were effective in the treat-ment of syphilis.

Although mercury compounds are now little used as antisyphilitics. many have found use as antiseptics (see Vol. 2, p. 83), fungicides (see VoL 6, pp. 986, 992, 993), and diuretics (see Vol. 5, p. 189) (see also *Mercury preparations)*. The use of organo-mercury compounds in synthesis is somewhat limited by their toxicity, the very vola-tile lower alkyls being especially dangerous.

Physical Properties

Table I lists some physical properties of a few typical alkyl- and al'yl mercury compounds.

Table

Compound	Physical appearance	M.p., °C	Density	Solubility	Other properties
$(CH_3)_2Hg$	Colourless liquid	b. 93-96	3.069	s. alc., ether	Deadly poison
CH_3HgCl	White crytls	170	4.063	—	Steam-volatile
$CH_3HgC_2H_3O_2$	Rhombic leaflets	142-143	—	s. acetic acid, ale.	Steam-volatile
$(C_2H_5)_2Hg$	Colourless liquid	b. 159	2.44	i. water, s. ether	Dedly pison; skin irritant
C_2H_5HgCl	Silvery leaflets	192.5	3.5	i. water; s. hot ether ether	More toxic than $(C_2H_5)_2Hg$
$(C_6H_5)_2Hg$	Small white needles	125-126	2.29-2.34	s. hot benzene, chloroform	Toxic
C_6H_5HgCl	White leaflets	251	—	sl. s. alc., ether, benzene	—
$C_6H_5HgC_2H_5O_2$	Small white prisms	149	—	s. benzene, alc., acetic acid	—
$C_6H_5HgNO_3$	White crystals	131-131.5	—	s. hot benzene	–

Reactions

As the least reactive organometallic compounds of Group IIB, organomercurials are not sensitive to active hydrogen compounds such as water, amines, and alcohols. Some mercury compounds can actually be purified by recrystal-lization from water. They are generally susceptible to strong acids only when these acids are hot and concentrated; strong bases usually do not react.

Unlike the compounds of cadmium and zinc, which are useful in preparing ketones by coupling with acid chlorides, mercury compounds do not undergo such a reaction as a rule. An exception to this case is the formation of thienyl methyl ketone by the reaction of thienylmercuric chloride with acetyl chloride. The following equations typify certain general reactions of organomercury compounds:

$$R_2Hg + HX \sim RHgX + RH$$

heat

$$C_2H.HgCl + (NH_4)2S \rightarrow (C_2H_5Hg)2S \rightarrow HgS + (C_2H_5)2Hg$$

$$R_2Hg + HgX_2 \rightarrow 2\ RHgX$$

$$R_2Hg + X_2 \rightarrow RX + RHgX$$

$$RHgX + X_2 \rightarrow RX + HgX_2$$

$$R_2Hg + BCl_3 \rightarrow RBCl_2$$

$$(C_3H_7)_2Hg + PCl_3 \rightarrow (C_3H_7)_3P$$

$$(CaH_7)_2Hg + AsCla + (CaH_7hAs$$

$$(C_3H_7)_2Hg + 3\ Na \rightarrow 2CH_3N_3 + N_3Hg$$

$$(C_2H_5)_2Hg + Be \rightarrow (C_2H_5)_2Be$$

$$(CH_3)_2Hg + Cd \rightarrow (CH_3)_2Cd$$
$$(C_2H_7)_2Hg + Al \rightarrow (C_3H_7)_3Al$$
$$(C_6H_5)_2Hg + Bi \rightarrow (C_6H_5)_3Bi$$

Preparation

Kumerous procedures have been reported for the synthesis of organic mercury compounds. Some of the more general methods include treating an alkyl or aryl halide (usually bromide or iodide) with sodium amalgam in the presence of ethyl acetate, the action of a Grignard reagent on mercuric halides, and finally the "mercuration" of an aromatic nucleus with mercuric acetate, nitrate, or chloride. These methods are summarized in the following equations:

$$CH_3I + NaHg \xrightarrow{CH,COOC_2H_5} (CH_3)_2Hg$$
$$C_6H_5Br + NaHg \xrightarrow{CH,COOC_2H_5} (CH_3)_2Hg$$
$$CH_3MgI + HgCb \longrightarrow (CH_3)_2Hg$$
$$CH_3MgI + HgCl_2 \text{ (excess)} \rightarrow CH_3HgCl$$
$$C_6H_5MgBr + HgCl_2 \rightarrow (C_6H_5)_2Hg$$
$$C_6H_5 + Hg(NO_3)_2 \rightarrow C_6H_5HgNO_3$$

A large number of medicinal and fungicidal organomercury compounds are available commercially. Among the simpler of these are ethylmercuric chloride, C_2H_6HgCl, phenylmercuric chloride, N.F. IX, C_6H_5HgCl, phenylmercuric acetate, $C_6H_5HgC_2H_30_2$, basic phenylmercuric nitrate, N.F. IX, N.N.R., approximately $C_6H_5HgNO_3.C_6H_6$-HgOH, phenylmercuric picrate tincture, N.N.R., phenylmercuric borate tincture, N.N.R., and other ethyl and phenyl mercurials, as well as a few analogous pyridyl mercurials.

Mercuration

Aromatic mercuration is a sufficiently general reaction to be con-sidered along with nitration, sulfonation, and halogenation as a typical substitution reaction. In this reaction an aromatic compound is treated with mercuric acetate, nitrate, or chloride, and a hydrogen atom of the aromatic nucleus is replaced by a mercuri-acid group. This reaction occurs so readily at times that polymercuration may take place.

Only recently the mercuration method has been improved to a point where in some cases it is the method of choice for preparing organomercurials. Thus benzene can now be mercurated in 80% yield with mercuric nitrate

Phenylmercuric Nitrate, $C_6H_5HgNO_3$

A mixture of 15 grams of reagent-grade mercuric nitrate, 5 grams of mercuric oxide, and 25 grams of Drierite (soluble anhydrous calcium sulfate) is stirred at 9800 r.p.m. in 200 ml. of thiophene-free benzene. The temperature of the reaction mixture is kept at 60°C. for ninety minutes. The system is protected from air by a rapid stream of carbon dioxide, evolved from a bottle of the solid and bubbled through concentrated sulfuric acid and then through benzene, the latter in order to replace the solvent lost from the reaction mixture as the gas passes through the flask. At the exit end the gases are pas!:ied through another flask that also contains solid carbon dioxide. At the end of the reaction the mixture is filtered. When the filtrate has cooled to room temperature, an equal volume of hexane is added. The phenylmercuric nitrate precipitate is then recovered by filtration on a Buchner funnel. The crude material melts at 126-128°C., and after two crystallizations from benzene it melts at 131-131.5°C. The recorded value is 130-132°C.

Acids and acid anhydrides can be mercurated by refluxing the sodium salt with aqueous mercuric acetate, or occasionally by heating the mercuric salt of the acid:

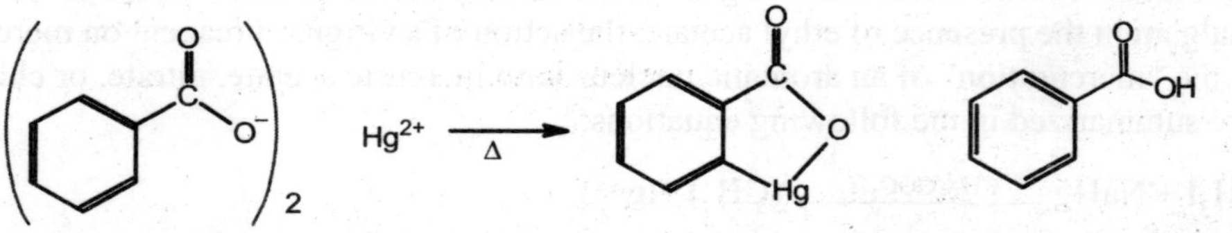

In the mercuration of acids as shown above, it is usually found that the mercury enters the aromatic nucleus ortho to the carboxyl group despite the strong meta direc-tion of this group. Anomalous orientation has been observed in numerous mercura-tion reactions.

Aromatic amines can be mercurated with extreme ease, often at or below room temperature:

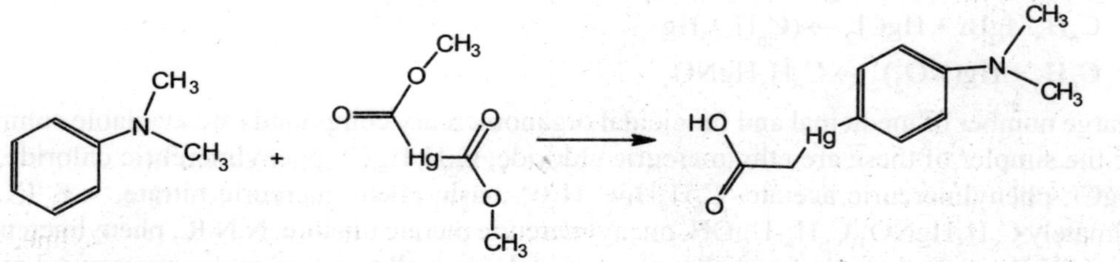

Heterocyclic compounds probably undergo mercuration more easily than any other aromatic type. One method of removing thiophene from benzene is to shake the mixture with a saturated aqueous solution of mercuric chloride. A practically quantitative reaction occurs:

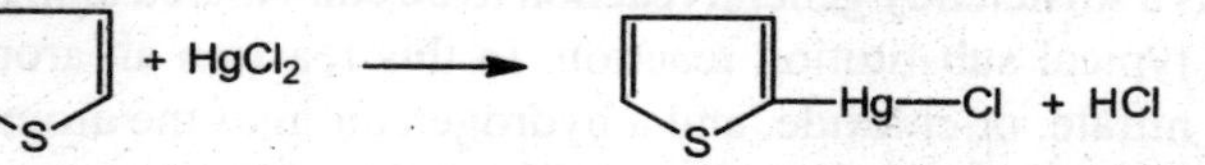

Furan, pyrrole, and pyridine derivatives can be mercurated similarly, and in most cases (except with pyridine derivatives, which give the 3-i~omer) the 2-position is attacked.

The mercuration of nitro compounds has been studied extensively, since the:;e compounds react quite readily to yield ortho mercurated products instead of the expected meta derivatives.

Phenols and phenol ethers are so easily mercurated that in some cases poly- mereuration occurs (especially with phenols) A typical preparation is that of o-ace-toxymercuriphenol:

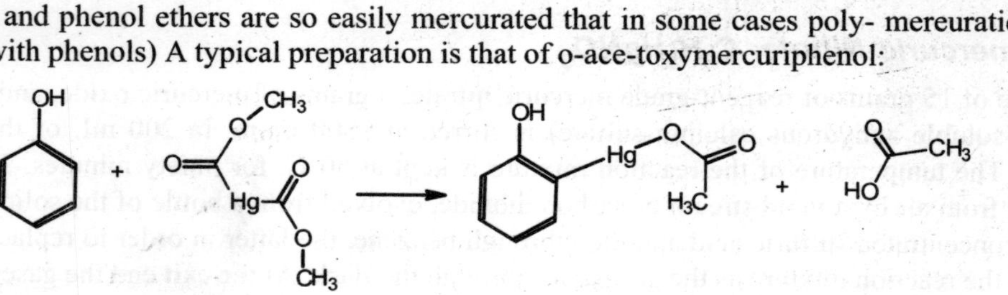

Phenol ethers are somewhat more difficult to mercurate than phenols.

Recently, Klapproth and Westheimer investigated the mechanism of aromatic mercuration and determined that when this reaction was carried out under truly ionic conditions (for example, mercuric perchlorate in perchloric acid solution), the orientations were normal for electrophilic substitution

Reactions with unionized mercuric acetate, however, showed much less strongly marked orientation effects.

Organo Aluminium Compounds

Aluminum

The first organoaluminum compound was reported by Hallwachs and Schafarik in 1859. They described the distillation of a liquid of the general Com-position $(C_2H_5)_3Al_2I_3$ obtained by treating ethyl iodide with aluminum. Cahours in 1860 investigated the product formed from methyl iodide and aluminum. In 1865 Buckton and Odling reported the preparation of aluminum alkyls by treating the corresponding mercury alkyls with aluminum. Following this early work was a series of papers by Spencer and Wallace and Leone describing the reaction of aluminum with a number of alkyl and aryl halides. However, none of these workers isolated or identified the compounds. It remained for Grignard in 1924 to separate and identity the mixtures obtained from these reactions.

Physical Properties

Table lists some physical properties of a few organo-aluminum compounds.

Compound	B.p. (m.p.), °C	Density	Refractive index	Other properties
$(CH_3)_3Al$	125.3 (dec.); 15 (m.p.)	0.752^{20}_{4}	1.432^{12}_{D}	Forms etherte readily; exists as dimer below 70°C.
$(CH_3)_3Al.(C_2H_5)_2O$	680_{15}	—	—	Ignites in air
$(C_2H_5)_3Al.(C_2H_5)_2O$	216-218; 112_{16}	0.8200^{17}_{4}	$1.4370^{17.4}_{D}$	—
$(CH_2)_2AlCl$	$83\text{-}84_{200}$	—	—	—
$(C_2H_5)_2AlI$	$118\text{-}120_{4\text{-}5}$	1.609^{27}_{4}	—	Dimeric; reacts violently with water
CH_3AlCl_2	72.7 (m.p.)	—	—	—
$C_2H_5AlI_2$	$158\text{-}160_{4}$	—	—	Dimeric; flammable in air; reacts with water
$(C_6H_5)_3Al$	Approx. 230 (m.p.)	—	—	Violently decomposed by water; easily soluble in benzene

Reactions

Aluminum alkyls and aryls tend to form addition complexes quite readily because of the ability of aluminum to expand its valence shell. Addition complexes with ether are especially common, and often are so stable that they can be distilled without decomposition.

Organoaluminum compounds either fume or are spontaneously flammable. They show the same general reactions as the Grignard reagents but at a distinctly slower rate. Aluminum displaces mercury from diarylmercury compounds more rapidly than zinc does. The low order of reactivity of organoaluminum compounds makes it appear very improbable that such compounds are formed as intermediates in the Friedel-Crafts reaction. The ethylaluminum iodides metal ate dibenzofuran in the 2- but not in the 4-position.

With ketones, organoaluminum compounds act simply as energetic condensing agents by removing water. Thus, when acetone is treated with ethylaluminum iodide, mesityl oxide, $CH_3COCH{:}C(CH_3)_2$, is formed. Acetophenone with triphenylaluminum produces dypnone, $C_6H_6COCH{:}C(CH_3)C_6H_5$.

Preparation

Generally speaking, three methods have been used successfully to prepare organoaluminum compounds. These are: reaction of alkyl or aryl halides with aluminum, reaction of organomercury compounds with aluminum, and reaction of two or more aluminum compounds to produce another. Perhaps the most comprehensive treatment of preparative methods for organoaluminum compounds is the work by Grosse and Mavity

The following description for the preparation of dimethylaluminum chloride and methylaluminum dichloride is typical of the first method:

Dimethylaluminum Chloride and Methylaluminum Dichloride. The reaction is conveniently carried out in an 800-ml. rotating autoclave equipped with a simple glass liner. It is important that the reaction be properly catalyzed at the start with either iodine, an aluminum halide, or an organoaluminum halide. Since the reaction is decidedly exothermic, cooling is advisable.

A typical charge consists of 38 grams of aluminum turnings and 38 grams of granular aluminum along with 1.4 grams of ethylaluminum bromide. The autoclave is cooled in ice, and methyl chloride is passed in from a charging cylinder in several portions until 120 grams has been added. The temperature rises sharply after each addition, 80 that cooling is necessary. The liquid products decanted from the bomb are distilled *in vacuo* through a Podbielniak column. A water-white liquid, (CH3hAICl, boiling at 83-84°C. at 200 mm., is obtained along with a white crystalline solid, $CH_3 - AlCl_2$. melting at 72.7°C.

Organo Tin Compounds

Tin

The first organic compound of this element, diethyltin, was prepared by Lowig in 1852. It is perhaps surprising that this first tin compound should contain bivalent tin, since, as a rule, such compounds are not as stable as those in which the tin is tetravalent. In fact, the dialkyl compounds show a tendency to revert to compounds in which tin has the higher valence. Alkyltin compounds containing different R groups are well known, and where the four groups are different, the compound can be separated into optical isomers, indicating the tetrahedral structure of the tin atom.

Physical Properties

Table III lists some physical properties of a few alkyl- and aryltin compounds

Compound	B.p. (m.p.) °C	Density	Refractive index	Solubility
$(CH_3)_4Sn$	78_{760}	$1.2913^{15.5}_{4}$	$1.5200^{25.5}_{D}$	i. H_2O; v.s. alc.
$(C_2H_5)_4$, Sn	175_{760}	$1.1988^{19.7}_{4}$	$1.4724^{19.7}_{D}$	i. H_2O; v.s. alc.
$(C_2H_5)_3Sn(n\text{-}C_3H_7)$	195_{764}	$1.168^{20.6}_{4}$	$1.4741^{17.5}_{D}$	—
$(C_6H_5)_4Sn$	225 (m.p.)	—	—	i. H_2O, alc.; v.s. hot chloroform, benzene
$(p\text{-}CH_2C_6H_4)_4Sn$	233 (m.p.)	—	—	—

In general, the tetraalkyl compounds have a rather pleasant, faintly ether like odour.

Reactions

If tetraalkyltin is treated with a halogen, one of the alkyl groups is removed:

$$R_4Sn + Cl_2 \rightarrow R_3SnCl + RCl$$

This is completely analogous to the reaction of a Grignard reagent or lithium reagent with halogen:

$$RLi + X_2 \rightarrow LiX + RX$$

Again, if a tetraalkyltin compound is treated with a mineral acid, cleavage occurs, similar to the reaction of the Grignard reagent with strong acids:

$$R_4Sn + HX \rightarrow R_3SnX + RH$$

$$RMgX + HX \rightarrow MgX_2 + RH$$

In mixed alkyl-arylstannanes the aryl groups are cleaved from the tin in preference to the alkyl groups. Thus, if trimethylphenylstannane is treated with acid, cleavage of the phenyl group occurs rather than of the methyls:

$$(CH_3)_3SnC_6H_5 + HX \rightarrow (CH_3)_3SnX + C_6H_6$$

Mercuric chloride effects a similar periferentialcleavage

$$(C_6H_5).SnC_2H_5 + HgCl \rightarrow C_6H_5HgCl + (C_6H_5)2Sn(C_2H_5)Cl$$

Perhaps the principal reactions of the halostannanes are hydrolysis and coupling with other more reactive organometallic compounds. Hydrolysis can be effected with most bases and leads to the corresponding hydroxy compounds:

$$R_3SnX + OH^- \rightarrow R_3SnOH + X^-$$

The hydroxystannanes are bases and therefore differ from the analogous silicon and germanium compounds, which have the properties of alcohols.

Preparation

The most successful of the numerous methods used to prepare organotin compounds find application also in the synthesis of other organometallic compounds of Group IV. These methods include the treatment of metallic tin with alkyl halides:

$$2CH_3I + Sn \xrightarrow[\text{tube}]{\text{Sealed}} (CH_3SnI_2)$$

This method is comparable to the so-called "direct method" now in commercial use for synthesizing organosilicon compounds. In this procedure a silicon-copper alloy is treated at elevated temperatures with methyl chloride and the principal product is dimethyldichlorosilane, $(CH_3)2SiCl_2$.

Another very satisfactory method for synthesizing tin compounds (applicable also to silicon and germanium compounds) involves treating tin tetrachloride with organo-zinc, -magnesium, or -lithium compounds:

$$SnCl_4 + 2\,(CH_3)_2Zn \rightarrow (CH_3).Sn + 2\,ZnCl_2$$

$$SnCl_4 + 4\,C_6H_5MgBr \rightarrow (C_6H_5)_4Sn + 2\,MgBr_2 + 2MgCl_2$$

$$SnCl_4 + 4\,(CH_3)_2NC_6H.Li \rightarrow (CH_3)_2NC_6H_5)_4Sn + 4\,LiCl$$

The catalytic decomposition of an aryldiazonium double salt with tin tetrachloride can also be used:

$$ArN_2Cl.SnCl_4 \rightarrow Ar_2SnCl_2 + N_2$$

Trialkyl- and -aryl tin halides can also be prepared by treating an R_4Sn compoun with halogen:

$$(CH_3)_4Sn + Cl_2 \xrightarrow{0^\circ\ dark} SnCl + CH_3Cl$$

$$(C_6H_5)_4 \xrightarrow{AlI_3} (C_6H_5)_3Sn + C_6H_5I$$

The following preparation of tetra-p-tolyltin by means of an organolithium reagent is typical:

Tetra-*p*-tolyltin, $(CH_3C_6H_6)_4Sn$. The calculated amount of p-bromotoluene in ether is added to fine shavings of lithium placed in absolute ether and stirred vigorously. To start the reaction a little of the bromide is added, and the mixture is heated a few minutes. The bromide is then added at a rate sufficient to keep the ether refluxing gently. After the addition is complete, the mixture is refluxed for one-half hour to complete the reaction, and the solution is filtered into a graduated dropping funnel. The solution, which is from 0.5 to 1.0 molar, is titrated to determine the content of lithium aryl. To a vigorously stirred solution containing an excess of p-tolyllithium is added 25 ml. of a benzene solution containing 1.3 grams of tin tetrachloride. The solution is stirred vigorously, refluxed for several hours, and then hydrolyzed with water. The yield of tetra-p-tolyltin is about 2.2grams (90%). The compound melts at 233°C.

Commercial Organotin Compounds

A few organotin compounds are on the market for use in stabilizing organic chlorine materials such as polyvinyl chloride resins. Tetraphenyltin, used as a stabilizer in chlorinated transformer oils, is sold as a powder in 150-lb. drums; compounds of the type R_2SnX_2 are available in technical grades (90% or above). Organotin compounds are also used as catalysts in the polymerization of silicones, and hexa-n-butylditin, $(C_4H_9)_3SnSn(C_4H_9)_3$ has been used as a synthetic lubricant.

The toxicological properties of the dialkyltin salts have not yet been fully investigated, but the vapors and dusts, and in some cases even the solid compounds, are known to be irritating to the skin and mucous membranes.

Di-n-butyltin oxide, $(C_4H_9)_2SnO_x$ is a white, amorphous powder which decomposes without melting when heated, and fires when exposed to flame. Its bulk density is 31.8 lb/cu.ft. It is soluble in hot butyl alcohol, amyl alcohol, glycerol, and organic esters, but insoluble in most other organic solvents and water.

Di-n-butyltin dichloride, $(C_4H_9)_2SnCl_2$, is a white, crystalline solid; m.p.,43°C.; vapour pressure at 100°C., 2 mm.; sp.gr., 1.36 refractive index,*1.499;* flash point, 335°F. It is insoluble in cold water, but is hydrolyzed on heating with water; it is soluble in organic solvents.

Di-n-butyltin diacetate, $(C_4H_9)_2(C_2H_3O_2)2$, is a clear, colorless liquid which decomposes on heating; f.p., 5-10°C.; sp.gr., 1.3125; *refractive index,* 1.482; flash point (Cleveland open cup), 290°F.; viscosity at 25°C., 18.2 centipoises. It is insoluble in water, but soluble in most organic solvents.

Di-n-butyltin dilaurate, $(C_4H_9)_2\ Sn(C_{12}H_{23}O_2)_2$, forms colourless, soft crystals melting at 27°C.; the liquid cannot be distilled at 10 mm.; sp.gr., 1.0525; *refractive index,* 1.470; flash point (Cleveland open cup), 455°F.; viscosity at 27°C., 42.9 centipoises. It is insoluble in water and methanol, but soluble in other organic solvents.

Di-n-butyltin maleate, $((C_4H_9)_2SnOOCCH: CHCOO)_x$, is an amorphous, white powder or glass-like resin; m.p., 110°C.; bulk density, 23.1Ib./cu.ft.; flash point (Cleveland open cup), 400°F. It is insoluble in water, but soluble in benzene, ethyl alcohol, and organic esters.

"***RS-31***" is a thio organotin compound (15.5% min, tin, 9.4-9.9% sulfur) which exists in a colourless to pale yellow liquid freezing below 4.5°C.; it cannot be distilled at 10 mm., and decomposes at 200°C.; sp.gr., 1.11325; *refractive index,* 1.504; flash point (Cleveland open cup), 350°F.; viscosity at 25°C.,33.2 centipoises. It is insoluble in water, but soluble in organic solvents.

Organo Antimony Compounds

The organoantimony compounds include most of the organoarsenic types and are named analogously, but no primary or secondary stibines are known, only tertiary, R_4SbOH. The stibonium hydroxides, R_4SLOH, like the arsonium hydroxides, are strong bases. The arylstibonic acids such as stibanilic acid, **p-$NH_2C_6H_4$-$SbO(OH)_2$**, and its derivatives have been studied to a considerable extent because of their pharmacological interest.

Physical Properties

The trialkylstibines are poisonous liquids, fuming and some-times igniting in air. Some physical constants of a few organoantimony compounds are given in Table.

Physical Constants of Organoantimony Compounds

Compound	B.p. (m.p.) °C	Density	Solubility
$(CH_3)_2Sb$	80.6	$1/523^{15}$	sl. s. H_2O; s. org. solvents
$(C_2H_5)_3Sb$	75_{17}	1.3244^{16}	—
$(C_6H_5)_3Sb$	50 (m.p.)	1.4343^{25} (1.4998)	i. H_2O; s. oerg. solvents; sl.s. alc.
$(CH_2)_2SbCl$	155-160	—	—
CH_3SbCl_2	$115–120_{60}$	—	—
$(C_6H_5)_3SbCl_2$	143 (m.p.)	—	s. xylene; hydrolyzed by aq. or lc. alkali

Reactions

Like the corresponding arsines, organic add halogen to form stable dihalides R_3SbX_2; oxygen to form oxides R_3SbO; Sulphur to form sulphides R_3SbS, etc. Although organoantimony compounds are apparently not active enough to add groups, they undergo metal-metal exchange reactions with organo-alkali compounds

$$(C_6H_5)_3Sb + 3n\text{-}C_4HgLi \rightarrow (n\text{-}C_4Hg)_3Sb + 3\,(C_6H_5)Li$$

n liquid ammonia, RzSbX compounds react promptly with lithium, sodium, potasium, calcium, and barium to give deeply colored RzSbM compounds, which react eadily with RX compounds:

$$R_2SbBr + 2\,Na \rightarrow R_2SbNa + NaBr$$

$$R_2SbM + R'X \rightarrow R_2R'Sb + MX$$

Preparation. The best general method for the preparation of trialkyl and triaryl antimony compounds is the reaction between the trihalide and the Grignard or organoithium reagent. The Bart reaction for the synthesis of arsonic acids is also applicable to the arsostibonic acids, $ArSbO(OH)_2$, but not to bismuthonic acids.

Triphenylstibine (triphenylantimony), $(C_6H_5)_4Sb$. In a 2-1. round-bottomed flask with a merury-sealed mechanical stirrer, reflux condenser, and separatory funnel placed 40 grams (1.6 moles) of magnesium turnings and 200 ml. of dry ether. To this is added 100 mi. of a mixture of 260 grams (1.6 moles) of dry bromobenzene and 800 ml. of dry ether. As soon as the reaction starts, 200 ml more of dry ether is added, and the remainder of the bromobenzene solution is added at such a rate to cause gentle refluxing. After all the bromobenzene has been added, a solution of 114 grams of freshly distilled antimony trichloride in 300 ml. of dry ether is added slowly. When this addition is complete, the mixture is heated on a steam bath for one hour longer. The mixture, when cool, is hydrolyzed with ice water and filtered, and the ether layer is separated. The aqueous layer is extracted twice with 200-ml. portions of ether. The combined ether extracts are dried, and been the solvent is removed, leaving 145-160 grams (82-90%) of crude triphenylstibine, m.p. 49°C.

Crystallization fro.. a petroleum ether (b.p. 40-50°C.) gives a white solid melting at 50°C.

Chelate

History

The term *chelate*, meaning claw was first applied in 1920 by Sir Gilbert T. Morgan and H. D. K. Drew, who stated: "The adjective chelate, derived from the great claw or *chele* (Greek) of the lobster or other crustaceans, is suggested for the caliper like groups which function as two associating units and fasten to the central atom so as to produce heterocyclic rings

Chelation is the binding or complexation of a bi- or multidentate ligand. These ligands, which are often organic compounds, are called chelants, chelators, chelating agents, or sequestering agent. The ligand forms a **chelate complex** with the substrate. The term is reserved for complexes in which the metal ion is bound to two or more atoms of the chelating agent, although the bonds may be any combination of coordination or ionic bonds.

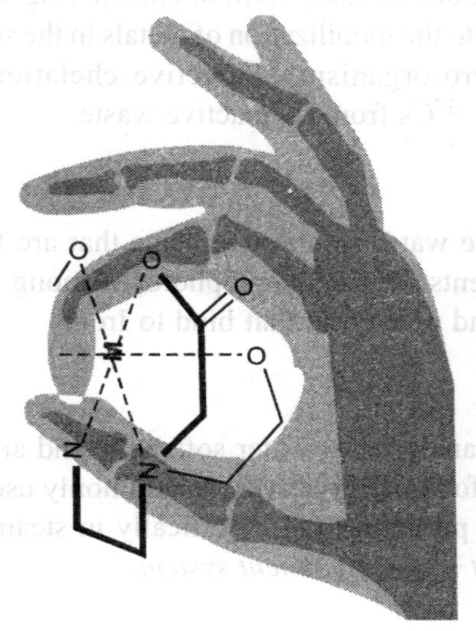

Fig. 37.1: The claw holding the organo-metallics forming a chelate

General

Relative to the aqua complexes, *e.g.* $[M(H_2O)_6]^{2+}$, the increased stability of a chelated complex, *e.g.* $[M(EDTA]^{2-}$ is called the chelate effect. Because chelating agents bind to metals through more than one coordination site, such ligands bind more tenaciously than unidentate ligands (like water). If a chelate were replaced by several monodentate ligands (such as water or ammonia), the total number of molecules would decrease, whereas if several monodentate ligands were replaced by a chelate, the number of free molecules increases. The effect is therefore entropic in that more sites are used by fewer ligands and this leaves more unbonded molecules: a total increase in the number of molecules in solution and a corresponding increase in entropy.

Chelation in Nature

Virtually all biochemicals exhibit the ability to dissolve metal cations. Thus proteins, polysaccharides, and polynucleic acids are excellent polydentate ligands for many of the metal ions. In addition to these adventitious chelators, several are produced to specifically bind certain metals. Such chelating agents include the porphyrin rings in hemoglobin or chlorophyll and the Fe^{3+}-chelating siderophores secreted by microorganisms. Histidine, malate and phytochelatin are typical chelators used by plants to avoid having poisonous metal ions in a free form.

In geology

In earth science, chemical weathering is attributed to organic chelating agents, *e.g.* peptides and sugars, that have the ability to solubilize the metal ions in minerals and rocks.[5] Most metal complexes in the environment and in nature are bound in some form of chelate ring, *e.g.* with "humic acid" or a protein. Thus, metal chelates are relevant to the mobilization of metals in the soil, the uptake and the accumulation of metals into plants and micro-organisms. Selective chelation of heavy metals is relevant to bioremediation, *e.g.* removal of ^{137}Cs from radioactive waste.

In microbiology

Many microbial species produce water-soluble pigments that are fluorescent under UV light. These pigments serve as chelating agents, termed siderophores. Among species of *Pseudomonas,* they are known to secrete pycocyanin and pyoverdin that bind to Iron.

Uses

Chelators are used in chemical analysis, as water softeners, and are ingredients in many commercial products such as shampoos and food preservatives. A commonly used synthetic chelator is EDTA. The term is used in water treatment programs and specifically in steam engineering, to describe a boiler water treatment system: *Chelant Water Treatment system.*

In medicine

Antibiotic drugs of the tetracycline family are chelators of Ca^{2+} and Mg^{2+} ions. Chelation therapy describes the use of chelating agents to detoxify poisonous metal agents such as mercury, arsenic, and lead by converting them to a chemically inert form that can be excreted without further interaction with the body. Chelation is also used as an unscientific treatment for autism or other conditions. There are no published peer review publications regarding the efficacy of chelation agents for the treatment of autism.

EDTA chelation can be a dangerous practice, especially when Na_2EDTA is prescribed rather than CaEDTA. The CDC reports that use of Na_2EDTA has resulted in fatalities due to hypocalcemia.

EDTA is also used in root canal treatment as a way to irrigate the canal. EDTA is used as a chelating agent to either soften the dentin facilitating access to the entire canal length and to remove the smear layer formed during instrumentation.

Gadolinium(III) chelates are often used as contrast agents in MRI scans.

Some Chelate Compounds

EDTA is a widely-used abbreviation for the chemical compound ethylenediaminetetraacetic acid (and many other names, see table). EDTA refers to the chelating agent with the formula $(HO_2CCH_2)_2NCH_2CH_2N(CH_2CO_2H)_2$. This amino acid is widely used to sequester di- and trivalent metal ions. EDTA binds to metals via four carboxylate and two amine groups. EDTA forms especially strong complexes with Mn(II), Cu(II), Fe(III), and Co(III).

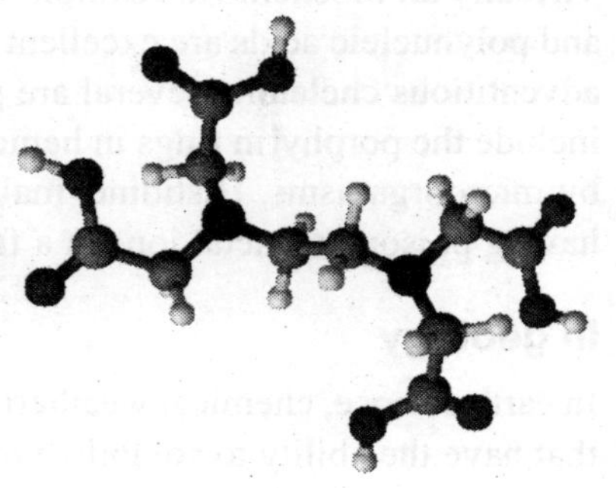

Synthesis

EDTA is mostly synthesised from 1,2-diaminoethane (ethylenediamine), formaldehyde (methanal), water and sodium cyanide. This yields the tetra sodium salt, which can be converted into the acidic forms by acidification. Pioneering work on the development of EDTA was undertaken by Gerold Schwarzenbach in the 1940's.

Popular vs. Chemical Nomenclature

To describe EDTA and its various protonated forms, chemists use a more cumbersome but more precise acronym that distinguishes between $EDTA^{4-}$, the conjugate base that is the ligand, and H_4EDTA, the precursor to that ligand.

Coordination Chemistry Principles

In coordination chemistry, H_4EDTA is a member of the aminocarboxylate family of ligands that includes imidodiacetic acid ("H_2IDA") and nitrilotriacetic acid ("H_3NTA"). More specialized relatives include N,N'-ethylenediaminediacetic acid ("H_2EDDA") and 1,2-diaminocyclohexane-N,N,N',N'-tetraacetic acid ("H_4CyDTA"). These ligands are all formally derived from the amino acid glycine.

H_4EDTA forms highly stable coordination compounds that are soluble in water. In these complexes, the ligand is usually either hexa- or pentadentate, $EDTA^{4-}$ or $HEDTA^{3-}$, respectively. Such complexes are chiral, and $[Co(EDTA)]^-$ has been resolved into enantiomers.

Uses

In 1999, the annual consumption of EDTA was equivalent to about 35,000 tons in Europe and 50,000 tons in the US. The most important uses are:

- **Industrial cleaning:** complexation of Ca^{2+} and Mg^{2+} ions, binding of heavy metals.
- **Detergents:** complexation of Ca^{2+} and Mg^{2+} (reduction of water hardness).
- **Photography:** use of Fe(III)EDTA as oxidizing agent.
- **Pulp and paper industry:** complexation of heavy metals during chlorine-free bleaching, stabilization of hydrogen peroxide.
- **Textile industry:** complexation of heavy metals, bleach stabilizer.
- **Agrochemicals:** Fe, Zn and Cu fertilizer, especially in calcareous soils.
- **Hydroponics:** iron-EDTA is used to solubilize iron in nutrient solutions.

More specialised uses of EDTA are:

- **Food**: added as preservative to prevent catalytic oxidation by metal ions or stabilizer and for iron fortification.
- **Approved by the FDA** as a preservative in packaged foods, vitamins, and baby food.
- **Personal care:** added to cosmetics to improve product stability.
- **Oil production:** added into the borehole to inhibit mineral precipitation.
- **Dairy and beverage industry:** cleaning milk stains from bottles.
- **Flue gas cleaning:** removal of NO_x.

- Dentistry as a root canal irrigant to remove organic and inorganic debris (smear layer).
- **Soft drinks** containing ascorbic acid and sodium benzoate, to mitigate formation of benzene (a carcinogen).
- **Recycling:** recovery of lead from used lead acid batteries.

Medicine:

- EDTA is used in chelation therapy for acute hypercalcemia, mercury poisoning and lead poisoning
- Combined with chromium, EDTA is used to evaluate kidney function. It is administered intravenously and its filtration into the urine is monitored. This method is considered the gold standard for evaluating glomerular filtration rate, Cr-EDTA's sole way out of the body is via glomerular filtration as it is not secreted or metabolised in any other way.
- Used as anticoagulant for blood samples
- In veterinary ophthalmology EDTA may be used as an anticollagenase to prevent the worsening of corneal ulcers in animals.
- Some laboratory studies also suggest that EDTA chelation may prevent collection of platelets ([or plaque] which can otherwise lead to formation of blood clots and prevent blood flow) on the walls of blood vessels [such as arteries]. These ideas are theoretical, however a major clinical study of the effects of EDTA on coronary arteries is currently (2008) proceeding

In laboratory science, EDTA is also used for:

- Scavenging metal ions: in biochemistry and molecular biology, ion depletion is commonly used to inactivate metal-dependent enzymes which could damage DNA or proteins
- Complexometric titrations.
- Buffer solutions.
- Determination of water hardness.
- EDTA may be used as a masking agent to remove a metal ion which would interfere with the analysis of a second metal ion present
- An anticoagulant in medical and laboratory equipment.
- A preservative (usually to enhance the action of another preservative such as benzalkonium chloride or thiomersal) in ocular preparations and eyedrops. See "les conservateurs en opthalmologie" Doctors Patrice Vo Tan and Yves lachkar, Librarie Médicale Théa.
- A titrant used to determine nickel concentration in an electroless nickel plating bath.
- In metallography to remove staining due to etchants. Metal oxides are removed by gently swabbing with EDTA and rinsing in water.
- In cell cultures EDTA is used as a chelating agent which binds to calcium and prevents joining of cadherins between cells, preventing cell clumping. (often used in cell culture control).

Toxicity

- EDTA has been found to be both cytotoxic and weakly genotoxic in laboratory animals. Oral exposures have been noted to cause reproductive and developmental effects.

Environmental Behaviour

EDTA coordinating a Cu^{2+} ion.

Widespread use of EDTA and its slow removal under many environmental conditions has led to its status as the most abundant anthropogenic compound in many European surface waters. River concentrations in Europe are reported as 10-100 μg/L, and lake concentrations are in the 1-10 μg/L range. EDTA concentrations in U.S. groundwater receiving wastewater effluent discharge have been reported at 1-72 μg/L, and EDTA was found to be an effective tracer for effluent, with higher concentrations of EDTA corresponding to a greater percentage of reclaimed water in drinking water production wells.

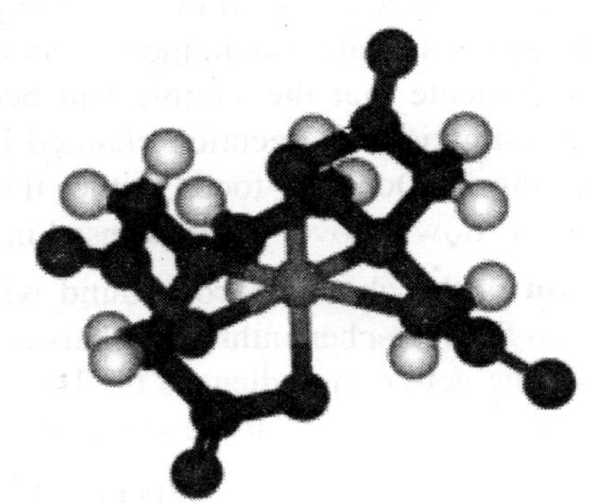

Fig. 37.2: EDTA coordinating a Cu^{2+} ion Forming an E.D.T.A. copper chelate

EDTA is not degraded or removed during conventional wastewater treatment. However, an adjustment of pH and sludge residence time can result in almost complete mineralization of EDTA. A variety of microorganisms have been isolated from water, soils, sediments and sludges that are able to completely mineralize EDTA as a sole source of carbon, nitrogen and energy.

Recalcitrant chelating agents such as EDTA are an environmental concern predominantly because of their persistence and strong metal chelating properties. The presence of chelating agents in high concentrations in wastewaters and surface waters has the potential to remobilize heavy metals from river sediments and treated sludges, although low and environmentally relevant concentrations seem to have only a very minor influence on metal solubility. Low concentrations of chelating agents may either stimulate or decrease plankton or algae growth, while high concentrations always inhibit activity. Chelating agents are nontoxic to many forms of life on acute exposure; the effects of longer-term low-level exposure are unknown. EDTA at elevated concentrations is toxic to bacteria due to chelation of metals in the outer membrane. EDTA ingestion at high concentrations by mammals changes excretion of metals and can affect cell membrane permeability.

Methods of Detection and Analysis

The most sensitive method of detecting and measuring EDTA in biological samples is selected-reaction-monitoring capillary-electrophoresis mass-spectrometry (abbreviation SRM-CE/MS) which has a detection limit of 7.3 ng/mL in human plasma and a quantitation limit of 15 ng/mL. This method works with sample volumes as small as ~7-8 nL.

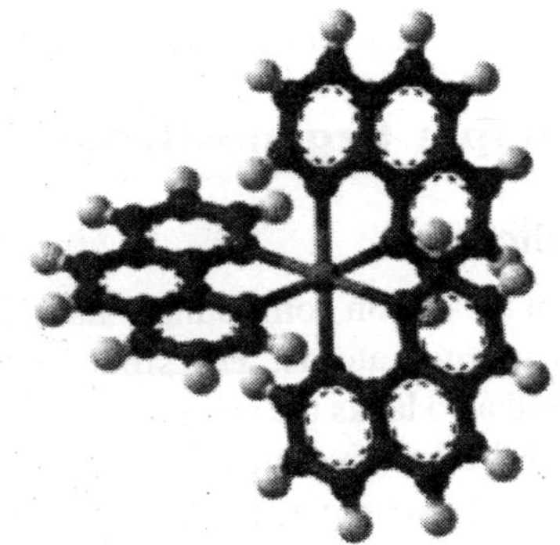

EDTA has also been measured in non-alcoholic beverages using high performance liquid chromatography (HPLC) which has a detection limit of 0.6 μg/mL and a quantitation limit of 2.0 μg/mL.

Forensics

- EDTA played a role in the O.J. Simpson trial when one of the blood samples collected from Simpson's estate was found to contain traces of the compound. This was used by the defense to indicate that the sample had been planted from one of the vials collected during the investigation. Prosecution claimed EDTA might have appeared in the sample as a result of eating McDonald's foods (either through bloodstream or, more likely, via contamination of blood flowing over the hand used in grabbing the food).

Ferroin is the chemical compound with the formula $[Fe(o\text{-phen})_3]SO_4$, where *o*-phen is an abbreviation for 1,10-phenanthroline. This coordination compound is used as an indicator in analytical chemistry. The active ingredient is the $[Fe(o\text{-phen})_3]^{2+}$ ion, which is the redox-active chromophore. Ferroin includes other salts of the same basic dication.

Its redox potential is +1,06 volts in 1M H_2SO_4.

Preparation of Reagent

A solution of 1.485 g 1,10-phenanthroline monohydrate is added to a solution of 695 mg $FeSO_4 \cdot 7H_2O$ in water, and the resulting red solution is diluted to 100 mL.

Silicon Organics

Introduction

Our earths crust is full of minerals and most of it is *siliceous*. On the other hand, the rule of carbon chemistry provides the utmost flexibility, variety and selectivity. The combination of these two elements was expected to provide something original. In this respect, the discoverers of silicones, Fredrick. S. Kipping and his co –workers were not disappointed.

The union of small hydrocarbon group with silicon – oxygen chains result in products with thermal stability, together with some unique properties, such as relative independence of certain physical properties of certain physical properties to temperate difference, for example, silicone oils change little in viscosity between temperature from 50–500°C.

Chemical inertness and lack of affinity for other materials are other characteristics of silicones. Silicone oils do not swell rubber nor are silicon rubbers swelled by ordinary oils and greases.

Simple Organosilicon Compounds

Silicones

Not all silicon compounds share the stability and inertness of those in which silicon atoms are connected by oxygen atoms. The silicon-hydrogen bond is a weak and highly reactive one. The silicon-silicon bond also lacks the stability of the carbon-carbon. Consequently, strictly silicon analogues of hydrocarbon and their derivatives are few and reactive. Replacement of one or more hydrogen atoms by increase stability as long as two or more silicon atoms remain on silicon atoms. Thus, silane, or silicane SiH_4 and its homologous react with many reagents rapidly in contrast to the behaviour of alkanes.

(1) $SiH_4 + HI \xrightarrow{80°C} H_3SiI + H_2O$

(2) $H_3SiI + HI \xrightarrow{AlI_3} H_2SiI_2 + H_2$

(3) $n - Si_2H_6 + 2nH_2 \rightarrow (H_2SiO)_n + 3nH_2$

Alkyl Chlorosilanes

The silicon organic compounds of primary interest as raw materials are alkylated silicon halides. These materials are produced mainly by two methods. One is the treatment of silicon tetrahalides with Grignard reagent, the other, direct reaction of elementary silicon with alkyl halides in the presence of copper. By either method mixture of all possible mono and polyalkyl derivatives are formed.

(4) $SiCl_4 + RMgCl \rightarrow R - SiCl_3 + MgCl_2$

(5) $Si + 2RX \xrightarrow{Cu} R_2SX_2$

(6) $R_2SiX_2 \leftrightarrows R_1SiX_2 + R_3SiX \rightarrow SiX_4 + R_4Si$

The ability of alkyl silicon halides to interchange halogen atoms and alkyl groups provides a means of modifying the ratio of products by introducing silicon tetrachloride or tetraalkylsilane and heating the mixture with a catalyst. The products can be separated by distillation, since the boiling points differ appreciably in most cases.

Table showing the boiling points of Alkylchlorosilanes

Formula	Name	Boiling Point °C
$SiCl_4$	Silicon tetrachloride	57.6
CH_3SiCl_3	Methyltrichlrosilane	64.6
$(CH_3)_2SiCl_2$	Dimethyldichlorosilane	69.4
$(CH_3)_4SiCl$	Trimethylchlorosilane	59
$(CH_3)_4Si$	Tetramethylsilane	26.6
$C_2H_5\ SiCl_3$	Ethyltrichlrosilane	100
$(C_2H_5)_2\ SiCl_2$	Diethyldichlrosilane	130.5
$(C_2H_5)_2\ SiCl$	Triethylchlrosilane	143.5

Carbon Functional Silicon Organics

The very inertness of hydrocarbon groups attached to silicon atoms that is attractive in silicones is a hindrance to introduction of groups substitution. Other method has to be derived. It is not suitable in most cases to have the functional group already present on the carbon atoms when the group is joined to the silicon atom, since the silicon halides and halogenosilanes that are starting materials or intermediates react readily with many functional groups.

Double bonds can be introduced into organosilicon compounds by using unsaturated halides in the reaction with silicon or by treatment of partly chlorinated silanes with acetylene the equation is shown below. The vinyl silanes show the addition reaction of double bonds connected to electron withdrawing atoms. This is just the reverse of the conjugative effect noted with halogen atoms.

$$2CH_2=CHCH_2+Si\xrightarrow{Cu}(CH_2=CH_2)_2SiCl$$

$$HC=CH+HISiCl_3\xrightarrow{Pt}CH_2CHSiCl_3$$

The silicon atom which can expand the valence level to as many as six pairs of electrons, with draws electrons conjugatively, releases them inductively. Orientation of groups going into the double bond is indicated in equations below. Other vinyl silicon compounds behave similarly.

$$CH_2=CHSi(OC_2H_5)_3+HBr\xrightarrow{P_2O_5}BrCH_2CH_2Si(C_2H_5)_3$$

$$CH_2=CHSi(OC_2H_5)_3+R-SH\rightarrow R-SCH_2CH_2Si(OC_2H_5)_3$$

$$CH_2=CHSi(OC_2H_5)_3+R_2NH\rightarrow R-NCH_2CH_2Si(OC_2H_5)_3$$

Halogenation can be accomplished by substitution, but special conditions are required. Bromination of methyl Chlorosilanes, for example, is best accomplished using bromine chloride.

$$(H_3C)_3SiCl + Br-Cl \longrightarrow BrCH_2Si(CH_3)_2Cl + HCl$$

An example of the unusual electronic influence of the silicon atom is its effect on chlorination in the benzene ring. The para-position is deactivated by the conjugative electron withdrawal but the ortho- and meta-

$$C_6H_5SiCl_3 + 4Cl_2 \xrightarrow[CCl_4]{AlCl_3} C_6HCl_4SiCl_3 + 4HCl$$

positions are activated by the inductive electron release.

A modified Wurtz-Fitting synthesis can be used to introduce certain carbon functional radicals into silanes. The example given shows use of another derivative as a protective group.

$$(H_3C)_3Si-Cl + HO-C_6H_4-Cl \xrightarrow{quinoline} (H_3C)_3SiO-C_6H_4-Cl + HCl$$

$$(H_3C)_3SiO-C_6H_4-Cl + 2Na + (H_3C)_3Si-Cl \longrightarrow (H_3C)_3Si-O-C_6H_4-Si(CH_3)_3 + 2NaCl$$

$$(H_3C)_3Si-O-C_6H_4-Si(CH_3)_3 + H_2O \longrightarrow (H_3C)_3Si-OH + HO-C_6H_4-Si(CH_3)_3$$

Grignard reagents can be prepared from chloromethylsilicon compounds this opens the way for the introduction of a wide variety of

$$(H_3C)_3Si-O-Si(CH_3)_2-CH_2Cl + Mg \xrightarrow{\text{diethyl ether}} (H_3C)_3Si-O-Si(CH_3)_2-CH_2-MgCl$$

The Oxo-process has also been used to prepare aldehydes from allyl silanes.

$$(H_3C)_3Si-CH_2-CH{=}CH_2 + CO + H_2 \xrightarrow[\Delta]{CO} (H_3C)_3Si-CH_2-CH_2-CH_2-CH{=}O$$

Reactions of Silicon Functions

The utility of Halogenosilanes is a result of the high reactivity of the silicon-halogen bond. This is due to the ability of the silicon atom to expand its valence level.

Halogenosilanes are readily hydrolysed to the hydroxy compounds and alcoholysed to a silicate ester.

$$(H_3C)_3Si-CH_2-CH_2-\overset{O}{\overset{\|}{C}}-OH + H_2O \xrightarrow[\text{Conc}]{H_2SO_4} 2CH_4 + HO-\underset{O}{\underset{\|}{C}}-CH_2-CH_2-SiH(CH_3)-O-Si(CH_3)_2-CH_2-CH_2-\overset{O}{\overset{\|}{C}}-OH$$

Silicones

The term silicone is applied to organo silicon compounds of the composition R_2SiO_3. However, it is apparent that siluicons do not in any way resemble ketones; their structures and properties are those of polyoxysilylenes (*i*) The term siloane has been accepted as the official designation of this type of compound. The usual numerical prefixes indicate the number of silcon atoms present, as in octoamethyltrisiloxane (*ii*) Silicones are prepared from appropriate mixtures of alkyl-chlosilanes by hydrolysis.

$$\left[-\overset{R}{\underset{R}{Si}}-O-\right] \quad (I)$$

$$H_3C-Si(CH_3)_2-O-Si(CH_3)_2-O-Si(CH_3)_2-CH_3 \quad (II)$$

Silicone Oil and Waxes

Treatment of mixture of dialkyldichlorosilanes and trialkylchlorosilanes with water gives linear siloxanes. Depending on their molecular weights, these compounds are oils and or waxy solids. The average molecular weight of the mixture can be controlled by the proportion of the compounds or components.

$$R_3SiCl + Cl{-}SiR_2{-}Cl + ClSiR_3 + (X + !)\,H_2O \longrightarrow R{-}Si(=O){-}SiR\;O^{+}{-}SiR_3 + (2n + 2)HCl$$

Silicone oils (polymerized siloxanes) are silicon analogues of carbon based organic compounds, and can form (relatively) long and complex molecules based on silicon rather than carbon. Chains are formed of alternating silicon-oxygen atoms (...Si-O-Si-O-Si...) or siloxane, rather than carbon atoms (...C-C-C-C...). Other species attach to the tetravalent silicon atoms, not to the divalent oxygen atoms which are fully committed to forming the siloxane chain. A typical example is polydimethylsiloxane, where two methyl groups attach to each silicon atom to form $(H_3C)[SiO(CH_3)_2]_nSi(CH_3)$. The carbon analogue would be an alkane, e.g. dimethylpropane C_5H_{12} or $(H_3C)[C(CH_3)_2](CH_3)$.

Silicone oils could be a basis for silicon-based organic life, but their more prosaic, primary uses are as lubricants or hydraulic fluids. They are excellent electrical insulators and, unlike their carbon analogues, are non flammable. Their temperature-stability and good heat-transfer characteristics make them widely used in laboratories for heating baths ("oil baths") placed on top of hotplate stirrers.

Some silicone oils such as simethicone are potent anti-foaming agents. They are used in industrial applications such as distillation or fermentation where excessive amounts of foam can be problematic. They are sometimes added to cooking oils to prevent excessive frothing during deep frying. Consumer products to control flatus (antiflatulents) often contain silicone oil. Silicone oils used as lubricants can be inadvertent defoamers (contaminants) in processes where foam is desired, such as in the manufacture of polyurethane foam. Silicone oils have been used as a vitreous fluid substitute to treat difficult case of retinal detachment, such as those complicated with proliferative vitreoretinopathy, giant retinal tears, and penetrating ocular trauma. Silicone oil is also one of the two main ingredients in Silly Putty, along with borax. Silicone oil plays a useful role in gas powered airsoft guns where it is used to lubricate the rubber gas seals in gas blowback guns without degrading them as carbon based oils would as well as lubricating the moving parts of the machines.

Siloxane

It is any chemical compound composed of units of the form R_2SiO, where R is a hydrogen atom or a hydrocarbon group. A siloxane has a branched or unbranched backbone of alternating silicon and oxygen atoms -Si-O-Si-O-, with side chains R attached to the silicon atoms.

The word *siloxane* is derived from the words **sil**icon, **ox**ygen, and alk**ane**.

Siloxanes can be found in products such as cosmetics, deodorant, water repelling windshield coating, food additives and some soaps. They occur in landfill gas and are being evaluated as alternatives to perchloroethylene for drycleaning. Perchloroethylene is widely considered environmentally undesirable.

Polymerized siloxanes with organic side chains (R "" H) are commonly known as silicones or as *polysiloxanes*. Representative examples are $[SiO(CH_3)_2]_n$ (**dimethylsiloxane**) and $[SiO(C_6H_5)_2]_n$ (**diphenylsiloxane**). These compounds can be viewed as a hybrid of both organic and inorganic compounds. The organic side chains confer hydrophobic properties while the -Si-O-Si-O- backbone is purely inorganic. Dimethylpolysiloxane is added to vegetable oil as an antifoaming agent. McDonalds uses this type of vegetable oil to cook certain products.

Siloxanes in Biogas

In internal combustion engines deposits on pistons and cylinder heads are extremely abrasive and causes damage to the internal components of the engine. Engines can require a complete overhaul at 5,000 h or less of operation. Deposits on the turbine of the turbocharger will eventually reduce the components efficiency.

Stirling engines are more resistant against siloxanes, though deposits on the tubes of the heat exchanger will reduce the efficiency.

Uses

Novel neopentasiloxane compounds of the general formula $Si[OSiR_2\,Q]_4$ wherein R denotes a monovalent hydrocarbon or halohydrocarbon group and Q denotes a group R or a group comprising a block of at least two oxyalkylene units and having a terminal group containing a reactive hydrogen atom, not more than one Q denoting an R group.

The invention also encompasses a film-forming composition comprising a curable polyurethane resin and the neopentasiloxane, and the use of the film-forming compositions to provide breathable (water vapour permeable) coatings on textile fabrics.

Oligomeric organosilicon compounds are disclosed of the formula I ##STR1## wherein R^1, R^2, R^3 independently of one another denote H, $(C_1 - C_4)$alkyl, $(C_1 - C_4)$ alkoxy, $(C_1 - C_4)$ haloalkoxy, $(C_1 - C_4)$ haloalkyl, phenyl, aryl or aralkyl and Z denotes an alkylidene radical having 0-6 carbon atoms, x can be a statistical average of 1-6 and n = 1-150 and . . . means that z can be bonded either to the one or the other C atom, and the particular free valency is occupied by a hydrogen.